Schedule an interview with Shelly Palmer
and get more information about the book.

www.shellypalmer.com/press

Digital Wisdom: Thought Leadership for a Connected World

Digital Wisdom: Thought Leadership for a Connected World
By Shelly Palmer

Shelly Palmer Digital Living Series

YORK HOUSE PRESS

Library of Congress Control Number: 2012955303
ISBN: 978-0-9855508-2-0

York House Press
1266 E. Main Street, Suite 700R
Stamford, CT 06902

Cover design, Annelise Pruitt
Book design, Lisa Lawlor

Dedication

To my wife, Debbie, who taught me what "connected" really means.

Praise for Digital Wisdom

"Shelly has created a compelling and comprehensive "How To" guide for 21st century digital leadership. Ignore at your own peril!!! In the New World Order of consumer engagement, everyone needs to be a digital marketer. Shelly has given us a highly practical guide to help navigate the waters to digital leadership."

— Joseph V. Tripodi, Chief Marketing & Commercial Officer, The Coca-Cola Company

"As analytics unlocks the power of data, marketers must immerse themselves in the connected world. In Digital Wisdom, Shelly Palmer explains how mastering the new digital realities can lead to extra degrees of freedom in your life and your business."

— Stephen Quinn, EVP & Chief Marketing Officer, Walmart

"Digital Wisdom delivers mightily on its promise ... and then some. Shelly Palmer's new book is informative, insightful and, best of all, genuinely inspiring — like Shelly."

— David Sable, Global CEO, Y&R

"Shelly has done a tremendous job showing us how to solve 21st century problems with 21st century solutions, whether you are a digital native or novice this is a must read."

— Keith Weed, Chief Marketing and Communication Officer, Unilever

"Once again Shelly has created a book that will make you think differently about life in the 21st century. It's a must read - not only for aspiring Digital Leaders - but for all Leaders!"

— Paul Mascarenas, Chief Technical Officer, Ford Motor Company

"Digital Wisdom is a well-written, easily readable, practical book for anyone who wants to take a leadership role in a connected world."

— Kristen O'Hara, Chief Marketing Officer, Time Warner Global Media Group

Contents

Acknowledgments

Gleeful thanks to Penelope Holt, the world's most thoughtful and caring publisher. Thanks also to Ellen Lohman for her eagle-eyed proofreading and editorial contributions, and a tip-of-the-hat to Dick Contino, designers Lisa Lawlor and Annelise Pruitt, and the entire team at York House Press.

Very special thanks to my sons, Brent and Jared, who spent many hours discussing these ideas with me. There is truly nothing that delights me more than learning, discussing and thinking about life, the universe and everything with them. A special shout out to Jared for helping me write the section on Truthiness in a Connected World.

Perhaps the most important thank you goes to my daughter, Alexis Zinberg, whose excellent leadership and remarkable project management skills made it possible to spend the time required to create this work.

Every father wants to brag about their children, and I am no different—Alexis, Brent and Jared ... you guys rock!

Kudos to the teams at Shelly Palmer Digital Leadership and Advanced Media Ventures Group LLC. You know how important you are, but I wanted to acknowledge your exceptional work here.

And finally, thanks to my wife, Debbie, my biggest supporter and my harshest, most honest and most constructive critic. This book could not have happened without you.

About the Author

Shelly Palmer is Fox 5 New York's On-air Tech Expert (WNYW-TV) and the host of Fox Television's monthly show *Shelly Palmer Digital Living*. He also hosts United Stations Radio Network's *Shelly Palmer Digital Living Daily*, a daily syndicated radio report that features insightful commentary and a unique insider's take on the biggest stories in technology, media and entertainment. He is managing director of Advanced Media Ventures Group, LLC, an industry-leading advisory and business development firm, and a member of the Executive Committee of the National Academy of Television Arts & Sciences, (the organization that bestows the coveted Emmy® Awards).

Along with his contributions to the advancement of television, Palmer is a pioneer in the field of Internet technologies. He is the patented inventor of the underlying technology for Enhanced Television used by programs such as ABC's *Who Wants to Be a Millionaire* and ESPN's *Monday Night Football*.

Mr. Palmer is well known for his Emmy® Award-nominate series of "Hi-Tech" Television specials created for Fox Television as well as his Emmy® Award-nominated series *Live Digital with Shelly Palmer*, created and produced for NBC Universal. He also created *Midnight Money Madness* for TBS and created and produced *HotPop*, a teen lifestyle and music show, for Starz/Encore. Mr. Palmer has directed and produced dozens of national television commercials and has also enjoyed a highly distinguished career as a composer and producer. He was the recipient of the American Society of Composers, Authors and Publishers' (ASCAP's) 12th Annual Film and Television Music Award for ABC's hit series *Spin City*. He was also recognized the following season in the category of "Most Performed Television Themes." Palmer's music credits include the theme songs for *Live with Regis & Kelly*, *Rivera Live* and *MSNBC* as well as the digital debut of the "real" cat singing the classic "Meow, Meow, Meow, Meow" for Meow Mix.

Palmer is a popular speaker and moderator at technology and media conferences hosted by industry organizations, like the Consumer Electronics Show (CES), the National Association of Broadcasters Convention (NAB), the National Show presented by the National Cable Television Association (NCTA), the National Association of

Television Program Executives (NATPE), Digital Signage Expo (DSE) and other major trade events. He is a guest lecturer at the MIT Media Lab, the Stern School of Business at New York University, the Columbia Institute for Tele-Information (CITI) at Columbia University, Tsinghua University, Rutgers Business School and other top-tier colleges and universities.

A graduate of New York University's Tisch School of the Arts, he is a regular technology commentator for CNN and CNNi and a weekly columnist for the Jack Myers Media Business Report and the Huffington Post. Palmer is the author of *Television Disrupted: The Transition from Network to Networked TV*, 2nd Edition (York House Press, 2008), the seminal book about the technological, economic and sociological forces that are changing everything, and *Overcoming the Digital Divide: How to Use Social Media and Digital Tools to Reinvent Yourself and Your Career* (York House Press, 2011). For more information, visit www.shellypalmer.com.

Why Aspiring Digital Leaders Should Read This Book

I have a lofty goal for this writing: I want it to inspire you to become an exceptional digital leader. To accomplish my mission, I am going to ask you to expand your vision. I want you to pretend that from a technology perspective, not only is everything possible, it already exists. This will remove the statements "That's science fiction" and "That won't happen for years" from our discourse.

I want you to have the courage to try new things and be willing to fail. As you know, courageous people get just as scared as you do ... they just power through the fear.

That said, you will also need to be responsible about innovation. Just because something can be done does not mean it should be done. Responsibility should not temper your curiosity or initiative. Great leadership is a delicate balance between vision, courage, innovation, character and the ultimate creative tool—imagination.

Knowledge may make you powerful, but experience makes you wise. To become an exceptional digital leader you will combine your quest for knowledge, your willingness to think differently and your experience.

How is an exceptional digital leader different from any other type of exceptional leader? An exceptional digital leader is digitally literate.

Some people confuse literacy with fluency. They are not the same. I do not need to speak French to know that an email I've received is in French. Although I am not fluent in French, as a leader I will recognize the language and acquire the services of a fluent French-speaker to translate the message. I am literate enough in foreign languages to recognize French when I see it. I even speak a few words of French, "bon voyage" and "hors d'oeuvres." But I need someone fluent in French to translate.

Would a digital leader just use Google Translate (translate.google.com) and not hire a live person to do the work? That is the subject of this book. When is it OK to use a digital tool to do a job, and when does it make sense to use an older technology (like a living, breathing translator)?

Learning to make these kinds of decisions is a dynamic process. It is also iterative. You will never be done growing, and you will never be done learning. I think that's the greatest thing about our time—technology is changing so quickly, we can actually feel it change!

To be a leader, you will need a first follower. Let's think about how you might position yourself to acquire one.

Introduction

Most senior executives know that it is rarely a good idea to use oneself as a focus group. While it is possible that your personal tastes and passions may be a perfect match for some segment of your target audience, it is very unlikely that it is a perfect match for your entire audience. This is such an obvious truth, it should go without saying.

But I guess it's not obvious enough.

Over the past few months I've noticed more and more business leaders citing personal examples during strategic discussions. This is understandable, because everyone now has a smartphone or a tablet or both, but it's an alarming trend.

I was working with a very large FMCG (fast moving consumer goods) client, and the CMO (chief marketing officer) said, "I have that app. It sucks. I never use it … ever." The app in question was a super-popular movie ticket app and the strategy session was about a program that would leverage moviegoers and movie enthusiasts.

Just for fun, I asked the CMO what about the app was off-putting. The answer blew me away, and I quote: "It always asks if it's OK to use my location. Can't the stupid app figure out where I am and know it's OK to use my location? … The app sucks!"

So … I found myself in a room full of self-described "digital marketers," most of whom didn't know anything about the current implementation of privacy regulations. And to make matters worse, those who did know were scared to say a word because they didn't want to correct the boss. Amusing, but not uncommon.

That said, I get paid to make sure this stuff doesn't happen, so I politely explained the difference between the iPhone's permission request and the app's capabilities. To everyone's delight, the CMO accepted the explanation as fact, but then said: "It doesn't matter if it's the app or this iOS thing, people won't use it … it sucks!" I'm sure you can imagine how much fun the rest of the meeting was.

A few years back, before iPhones, before Android phones, before iPads, CMOs knew what they didn't know, and most of the great ones relied on market research and empirical data to set the stage for key strategic decisions. You could argue that the best marketers relied on their gut instincts, but I would push back and say that even the most emotional and artistic CMOs worked from a brand brief.

However, in the past 18 months or so, there has been a remarkable (and somewhat disturbing) trend toward the concept of "digital marketer," as opposed to just plain "marketer," who happens to use digital tools to help achieve marketing objectives. This unfortunate naming convention is in no way limited to the marketing department. Adding the word "digital" to just about any title has become a worrisome trend.

Now, while I've never heard any chief marketers announce which TV shows they personally enjoy when describing a TV buy, I've yet to have a conversation about social media, apps or online advertising that didn't include highly opinionated bias born of personal technology use. This is intrinsically bad, but not all bad. The good news is that from the Arab Spring to Occupy Wall Street to social networks, everyone is getting in the game.

People—consumers and marketers alike—are eagerly using technology to create meaning in their lives, and there is a growing understanding about how things like wireless networks and smart devices can change the world. The bad news is that everyone has such an emotional connection to their technology that people, including business leaders, are forgetting that everyone's experience is different.

There is no reason or rationale for formulating a worldview through the lens of your preferred technology and projecting it upon your target audience. If it sounded stupid when you read the previous sentence, it's even stupider when you enact it without realizing you're doing it.

The curse of personal experience is something business leaders have been successfully guarding against for years. Perhaps it's time to redouble our efforts to prevent "focus groups of one" from having undue influence. A little vigilance might prevent you from making serious tactical mistakes, such as killing partnerships with wildly popular, highly targeted apps just because you think they suck.

That said, this book is about digital leadership. It's a new twist on an old skill set. Digital leadership is a mindset, a discipline, a philosophy and a religion all rolled into one.

If you are guilty of imposing your inexperienced digital worldview on your business, you're not alone. The good news is you're reading this book and you are thinking about how to succeed in a connected world.

I don't have all the answers. In fact, I don't even have all the questions. But if I've done my job, the following collection of topics and ideas will empower you to think about how our connected world is evolving and how you can both lead and succeed.

Connected and Not Connected

Two Kinds of People and Two Kinds of Devices

Right now there are two kinds of people and two kinds of devices: connected and not connected. This may sound obvious or trivial; I assure you it is not. Contrary to popular belief, all people below the age of 25 are not "connected," and all people over the age of 65 are not "not connected."

We live in a world where the adoption of technology is lumpy and non-homogenous. How much technology you have been exposed to is a huge factor in determining where you are on the technology adoption and technological proficiency curve.

In my experience, people under the age of 25 and over the age of 45 are the most digitally skilled. Why? If you're under 25, you were born into an all-digital world. If you're over 45 and have kids, you were forced to learn a fair amount of digital skills to communicate with them, and your kids have been there to help you learn.

Often, the least technologically skilled people I meet are in their late 20s to early 40s. The people who make up this cohort were not born into a digital world and do not have kids old enough to help them, and when the rest of us were learning to use digital tools, this group had the "big job" and no time (or need) to become technologically savvy.

In truth, the reasons for the lumpy distribution of connected and not connected people and devices do not matter. What matters is that this is a fact of life in the 21st century, and every business leader will have to pay close attention to both connected and not connected people and devices in our evolving connected world.

Need proof? Credit cards and online payment systems have not created a cashless society. According to my friends at JPMorgan Chase Bank, 95 percent of all retail transactions are still done in cash. That's not my personal experience; I'll use a credit card to buy a cup of coffee. But the data do not lie.

Email and scanners were going to give us the paperless office. Nope. There are piles of paper everywhere. We scan everything in my office, but we still go through enough paper to start our own recycling business.

As fast as the world is changing—and I am going to spend some serious time trying to convince you that it is changing much faster than you think it is—there are some types of transactions that will always take place the way they always have.

The sad news is that you are going to be required to have centers of excellence simultaneously in both the connected and the not connected worlds in order to be considered a successful digital leader.

Stats

For obvious reasons, I am not a big fan of publishing statistics in books. But the trends are easy to identify:

1. There will be more smartphones and tablets deployed tomorrow than there are today.

2. The cost of digital file storage will be lower tomorrow than it is today.

3. The cost of computer power will be lower tomorrow than it is today.

4. There will be more people connected to digital networks tomorrow than there are today.

As you can see, there is a pattern—technology, right along with connectivity, is getting more powerful and less expensive. According to Intel, at the beginning of 2012 there were approximately 2 billion people connected to the Internet. By 2015, it expects the number to be over 3 billion, and remarkably, by 2020, Intel says there will be 4 billion people connected to the Internet using its chips. Cisco says that by 2020 there will be 15 billion connected devices in the world. Even if these numbers are not accurate, they are still important because the trend is clear: By 2020, there will be more connected people and connected devices in the world, as many as half the global population—but it still won't be everybody.

What does this mean to you as a digital leader?

Metamerica: Evolving the Governance of a Digital Democracy

Dateline New York: April 3, 2017—With 28 of the nation's 80 major data centers totally down and the rest just holding on, over 75 million people who rely on cloud-

based computer applications and storage have been completely crippled by a loss of data that is unprecedented. It has caused the biggest financial crisis since 2008. Hundreds of software-as-services companies have lost all of their data and the economic impact will easily reach $10 trillion of combined losses.

The White House issued a statement today urging people to stay calm. The Federal Data Depository Corporation is now saying that some people will have access to some of their data as early as next week. "Health records are job one," said the HHS director in a joint statement. "We have back-ups for everyone's dicom files and most of everyone's MRIs and about 500 million x-ray files on near-line storage. It will just take time to reload the data."

The FCC issued a similar statement promising that the 20 billion hours of user-uploaded content stored in the cloud since 2009 and the 500 million hours of professionally produced on-demand content may be available by the end of the year.

Sadly there is no word on when the Microsoft Office Cloud or the Adobe Cloud will be back online. And to the delight of most citizens, the IRS has lost almost every tax return for the past four years. There is one bit of good news: Verizon's and AT&T's LTE network can support VoIP calls, so at least some of the 165 million households and businesses without voice communication will have a work-around in the next few weeks.

How did it happen?

```
010101101101001011010010110100101101
010101001010110110110100101101001011 01
010101001010101101010100101010110100 10
101001011010010100101010110101010100 10
010010110101010010101011010010110101 01
101010110101010010101011010101001011 01
010010001001010101100100100010100101 01
001001001010101110110110110100110101 01
010001011010101010010111101101011010 11
010010110100101101010101101001011010 01
```

In Metamerica (the data that describe America), we are all a bunch of ones and zeros. Everything about us—our tax ID numbers, social security number, credit card numbers, contact information, financial data, medical history, entertainment preferences, computer software, the content we own – everything we are is described by ones and zeros on a magnetic or optical storage device somewhere.

Even today, we live in a world where metadata (data that describe other data) are more important than the actual data. For example: What good are the data on your iPod without the directory that tells you what songs and videos are available and where the computer should go to get them?

Similarly, as we enter the super-digital age, we (all of us) will be completely described by the metadata stored in Metamerica. It is the virtual country that describes our physical country, and to be perfectly frank, Metamerica is more valuable and more vulnerable than physical America.

The advent of cloud storage, cloud computing and centralized data warehousing is more than a trend; it is a paradigm shift toward the efficiencies and fiscal benefits of storing and manipulating the information in our world digitally.

Why pay for a group of IT technicians to patch servers and tend to racks of computers when you can plug into a computing and storage cloud and only pay for the services you use? It is identical to how America transitioned from generating its own electricity to purchasing electricity from the utility grid. The economics are too powerful for anything to stop this transition. It will be complete in a very few years. Want to see it today? Go to Google Apps for Business, salesforce.com, Yahoo!, Facebook—the list of software as services and subscription software and cloud storage companies is getting longer every day.

Now, we live in a world where everything about everyone is stored somewhere. The world is evolving toward a place where access to these data is going to empower all kinds of things, good and bad. This is not so much a privacy issue (although you could label it that way) as a governance issue. What should the structure of the government of Metamerica look like? There are no geographic restrictions, no towns, no counties, no states, no country—there are just Metamericans that describe Americans. Because Metamericans are data, they can be sorted any way an interested party might deem reasonable: by community of interest, by behavioral preferences, by medical issues, by DNA sample, by ... you name it. Metamericans are data, and data can be mined, analyzed and correlated in every imaginable way.

Is your DNA covered by copyright law? Are your media or television viewing habits? Are your medical records going to be available on a local storage device, regional server, your doctor's LAN or your HMO's WAN or at the fictitious, but reasonable to imagine, Federal Data Depository Corporation? If every bit of a Metamerican (pun intended) is data, which bits are protected, which are protectable, which are private, which are semiprivate and which are public? How many nontechnical people with digital medical records will die because there were no paper records of their medical condition or history and they could not afford, or didn't know how, to back up the data themselves?

When my father was transferred by ambulance to a hospital from another medical facility in an emergency situation, the staff forgot to give his paper records to the EMS technician who was transferring him. I got to the hospital an hour after he did, to find staff completely mistreating his illness because their broadband connection was down and they didn't have any paper records to base his treatment on. I had stopped by the other medical facility and picked up his charts, but what if I had been unable to do so?

To be honest, I'm not too worried about a terrorist attack with a dirty bomb or any other weapon of mass destruction. What I'm worried about is how we are very quickly becoming dependent on explosive amounts of centralized data and how unprotected, and unprotectable, those data are. Want to create a financial crisis that will make the recession of 2008 look like a trip to Disney World? Take out the data warehouse of the IRS, a few banks, a couple of credit reporting agencies, Google, Amazon and Microsoft. If it happened today, it would be hard to recover from. If it happens in 2017, chances are we will not be able to recover. These are just a few of the myriad issues we are going to have to think about as Metamerica and America evolve together.

How could the Great Data Crash of 2017 occur? Maybe it will be cyber-terrorism. Maybe it will be an incredibly robust computer virus. Maybe it will be a solar coronal mass ejection that is too powerful for our atmosphere to block. Sadly, there are about a zillion ways for it to happen and only one way to avoid it.

What we need is a cross-industry task force that would include experts from the military, information technologies, healthcare, legal, telecommunications, consumer electronics, entertainment, ethics and government to get together and have a serious, Socratic debate on the appropriate way to govern Metamerica. This is a nontrivial problem; it's about our economic sovereignty and our national security. To begin to solve it, we need to deal with the bigger question: How can analog leaders possibly govern a digital constituency?

Analog Leaders and Digital Workers

One of the most significant problems during the transition from the Industrial to the Information Age is the coexistence of analog leaders and digital workers. You can also think of this as analog generals and digital soldiers, analog doctors and digital patients, analog teachers and digital students ... no matter where you look, it is a significant problem.

Much has been written about the way Born Digitals (kid born after 1989 who have only lived in a digital world) are different from the previous generation. Some of it makes for very interesting reading, but for us, the lesson is already clear: We need to become digital leaders! We're going to start our journey to digital wisdom and leadership by understanding some of the fundamental realities of the Information Age. And it all starts with the rate of change.

The World Is Changing Faster than You Think

The Law of Accelerating Returns

In his book *The Singularity is Near*, Ray Kurzweil offers a thesis he calls "the law of accelerating returns." The math is simply summed up in Ray's famous quote, "We won't experience 100 years of progress in the 21st century—it will be more like 20,000 years of progress (at today's rate) because the pace of technological change is exponential."

Exponential growth is very hard for us to understand from our everyday experiences. We don't live in a world where things "feel" exponential to us. There are lots of great brainteasers to help us grasp exponential rates of change. There's the aphorism about the lily pond where the lily pads grow exponentially and will cover and kill all the life in the pond in 30 days. For most of the month, the plant growth seems small and contained. The associated question asks, "On what day will the lilies cover half the pond?" The answer is day 29, which leaves only one day to save it.

There's another story, possibly more famous, about putting one grain of rice on the first square of a chessboard and doubling it on each of the subsequent squares. The math is simple: 2^{n-1} grains of rice up to the nth square. So you need two grains on the second square, four grains on the third, eight grains on the fourth and on and on. You'll need more than 1 million grains of rice on the twenty-first square and more than 1 trillion (a million million) grains on the forty-first square, and by the time you get to the last square of the chessboard, there isn't enough rice in the world to accomplish the task.

Exponential growth is not something human beings naturally understand from common experience. That said, the evidence for the exponential rate of technological change is overwhelming and happening as you read this.

Moore's Law

According to Wikipedia, "Moore's law is the observation that over the history of computing hardware, the number of transistors on integrated circuits doubles approximately every two years. The period often quoted as "18 months" is due to Intel executive David House, who predicted that period for a doubling in chip performance (being a combination of the effect of more transistors and their being faster).[1]

Morre's law is named after Intel co-founder Gordeon E. Moore, who described the trend in his 1965 paper.[2][3][4] The paper noted that the number of components in integrated circuits had doubled every year from the invention of the integrated circuit in 1958 until 1965 and predicted that the trend would continue "for at least ten years."[5] His prediction has proven to be uncannily accurate, in part because the law is now used in the semiconductor industry to guide long-term planning and to set targets for research and development [6]"

While Wikipedia is not always the most accurate source for historical information, this is pretty much the accepted description of Moore's Law. Keep this in mind as we look at another observed law that will help us understand the speed and implications of the transition from the Industrial to the Information Age.

Metcalfe's Law

To continue with Wikipedia entries, "**Metcalfe's law** states that the value of a telecommunications network is proportional to the square of the number of connected users of the system (n^2). First formulated in this form by George Gilder in 1993,[1] and attributed to Robert Metcalfe in regard to Ethernet Metcalfe's law was originally presented, circa 1980, not in terms of users, but rather of 'compatible communicating devices' (for example, fax machines, telephones, etc.)[2] Only more recently, with the launch of the Internet, did this law carry over to users and networks, as its original intent was to describe Ethernet purchases and connections.[3] The law is also very much related to economics and business management, especially with competitive companies looking to merge with one another.

Metcalfe's law characterizes many of the networking effects of communication technologies and networks such as the Internet, social networking, and the World Wide Web. Former Chairman of the U.S. Federal Communications Commission Reed Hundt said that this law gives the most understanding to the workings of the Internet.[4] Metcalfe's Law is related to the fact that the number of unique connections in a network of a number of nodes (n) can be expressed mathematically as the triangular number n(n - 1)/2, which is proportional to n2 asymptotically.

The law has often been illustrated using the example of fax machines: a single fax machine is useless, but the value of every fax machine increases with the total number of fax machines in the network, because the total number of people with whom each user may send and receive documents increases.[5] Likewise, in social networks, the greater number of users with the service, the more valuable the service becomes to the community."

Statistics and Projections

As I've said, I'm not a fan of publishing statistics in a book. It dates the writing. However, in order to gain an understanding of our future, we need to acknowledge the most current stats and projections.

Again, according to Intel Corporation, there are over 2 billion people connected to the Internet. By 2015, there will be 3 billion, and by 2020, Intel projects that 4 billion people will be online worldwide. And according to Cisco Systems, by 2015, there will be over 15 billion connected devices worldwide.

Kurzweil, Moore and Metcalfe

When you take Ray Kurzweil's law of accelerating returns and put it in the context of Moore's law and Metcalfe's law, a remarkable timeline appears. It's an exponential curve that tells us that a vast amount of computer power, network power and vocational capability are all rapidly trending upward. But what does this mean?

Technology Is Meaningless Unless It Changes the Way We Behave

"Technology is meaningless unless it changes the way we behave" is a great filter that we can all use to determine what bright, new, shiny objects are important and which ones can be ignored. Is a new app a paradigm shift or a parlor trick? How about that new computer that has no storage capacity but connects to the cloud? Just ask yourself, "Will this new (fill in the blank) change the way people behave?" If you think it will, then pay attention to it. If you don't think it will, then you can probably ignore it.

So ... what happens when you ask this question about the curve we just created ... the nexus of the law of accelerating returns, Moore's law and Metcalfe's law?

It's easy to see how the nexus of these three observed laws will empower consumers and directly cause a dramatic behavior change. From enhanced showrooming (the behavior consumers exhibit when they use their mobile devices to check prices in a physical retail location, then purchase the item online for a lower price) to self-driving cars to endo-digitally enhanced human beings, this exponential increase in the network effect, the speed of computer power and the omnipresence of wireless access is a game changer.

However, the transition will not be smooth, nor will everyone profit from it. Powerful business, government and social forces are all interacting, and in most cases fighting as hard as they can to maintain the status quo.

The Information Super Toll Road: An Intended Consequence of Net Neutrality

What is Net Neutrality? On September 21, 2009, Federal Communications Commission (FCC) Chairman Julius Genachowski outlined the concrete actions he believed the commission would have to take to preserve a free and open Internet. He said, "The Internet is an extraordinary platform for innovation, job creation, investment, and opportunity. It has unleashed the potential of entrepreneurs and enabled the launch and growth of small businesses across America. It is vital that we safeguard the free and open Internet." The commissioner presented six principles that we might use to craft these new rules:

1. Consumers are entitled to access whatever lawful Internet content they want.

2. Consumers are entitled to run whatever applications and services they want, subject to the needs of law enforcement.

3. Consumers can connect to networks whatever legal devices they want, so long as they do not harm them.

4. Consumers are entitled to competition between networks, applications, services and content providers.

5. Service providers are not allowed to discriminate between applications, services and content outside of reasonable network management.

6. Service providers must be transparent about the network management practices they use.

This is the FCC's proposal. However, the specifics of the Net Neutrality fight and the proposed guidelines (which will ultimately become rules) are all a smoke screen. Let's talk about how the future is likely to unfold if a reasonable agreement is not reached.

There's an ongoing battle between Level 3 and Comcast. It's a good example of the kind of issues that are intended consequences of the Net Neutrality debate. Netflix uses Level 3 to deliver its videos to customers. It pays Level 3 for bandwidth. If you are a Comcast customer, you pay Comcast for bandwidth. In theory, Level 3 is getting paid, and Comcast is getting paid and everything should be fine.

However, in order for your Netflix movie to arrive at your Comcast-connected home, the bits have to pass from Level 3 to Comcast. If you're Comcast, this is an excellent place to put a tollbooth.

Until it became an issue, Level 3 and Comcast had an arrangement that allowed each company to send bits bidirectionally. The arrangement was made back when both companies sent about the same amount of bits to each other. But now that Level 3 is sending more bits through Comcast than Comcast is sending through Level 3, Comcast wants to be paid. This is the nature of the current battle in its simplest terms. But, like I said, it's a smoke screen.

The cliché description for the public Internet (courtesy of Al Gore) is the "Information Superhighway." It's a reasonable metaphor for the way information travels around the Internet. Even engineers like to call bunches of bits getting from place to place "traffic."

In the physical travel world, you can get from place to place several different ways. You can walk, ride a bike, take a car, take a bus, take a train or fly. Of course, while some places are accessible only by air, we all know that remote locations very often require us to use multiple modes of transportation. Now, imagine that people are bits.

The physical transportation world also has a fairly well-defined class structure. It is segmented with modes of travel that efficiently meet the needs of each constituency—and it is economically segregated. You are about as likely to find an Upper East Side socialite in the lounge at the Port Authority bus terminal waiting for a bus to South Carolina as you are to find a single mother of six on welfare in the Admiral's Club at JFK Terminal 8, waiting for her first-class seat to the Vineyard. It happens, but not often. Most of the time, modes of travel—air, train, bus, car, bike, feet—are a function of economic class, means and emergent need. So to keep the metaphor going, imagine that Comcast is Cathay Pacific Airlines and Time Warner Cable's basic broadband service is Amtrak and Verizon's low-end DSL service is Greyhound.

How would we expect the economic landscape to look in a world where, instead of one Information Superhighway, we'd have web of public and privately owned Information Super Toll Roads? Would we expect people who could only afford Greyhound bus service to do business with companies in Europe or Asia? Would we expect people who could only afford Amtrak train service to compete with people who could deliver merchandise overnight via air? Would organizations that own toll roads make it just a little too expensive to compete with them? Would organizations that own airlines charge competitors for extra bags and bigger seats? Keep asking travel questions; they all apply!

It is hard to be optimistic about a future world where there is a low-powered free and open Internet and a web of private toll roads owned by nongovernmental organizations that inherently compete with their customers. The specter of such a world bodes ill for innovation, entrepreneurship and in some ways even the doing of connected life.

This line of thinking raises the question: Will a plurality of Internet underclasses evolve? Want a current day analog? Look at the prepaid mobile phone business. It's huge, and so unstructured that even the service providers don't know who is using their products or how they are using them.

Back in the day, phone companies charged us for making calls, but receiving a call was free. This was a function of technology, not desire. As soon as cell phones hit the market, we started paying for time used (bandwidth) both coming and going. What's happening with Net Neutrality is a fight over exactly the same issue. Now, we pay for the bandwidth we use on one end. If this goes the wrong way, we (you and I) will pay for bandwidth both coming and going. On a personal level, this is not onerous. However, at the enterprise level, if we were to govern America for the best possible GDP (as opposed to governing for corporate profits), it has the potential to be a huge problem.

I am NOT advocating any government involvement with the Internet. I think government has proven that it has no business being in any business. However, this is not a debate you can leave to others. Get your elected officials on the phone. Take a few minutes to learn about the issue. We have to step up and become architects of our digital future. Become part of the solution. America needs you!

Will Tiered Pricing Lead to "ProxyNets" and Destroy the Universe?

From Google News: June 11, 2015 Washington, D.C.—Representatives from AT&T, Verizon and Comcast were up on The Hill today trying to persuade Congress to take aggressive action against the millions of ProxyNets that have popped up around the globe. With two of the major content companies in bankruptcy and the others on the brink of ruin, experts are not sure if any type of law protecting intellectual property can be practically enforced. You can't shut down a network that you can't find.

ProxyNet? Well, I needed a name for a universe-destroying technology that I am absolutely sure will evolve in the next couple of years. ProxyNet seems to fill the bill. Here's how it will transpire.

Wired and wireless carriers will have enacted strict tiered pricing. Apple will have just introduced yet another amazing device that allows the joyful, gluttonous consumption of rich media over a closed network. With the advent of a zillion apps and competitive devices that will aid and abet gluttonous media consumption, people will start to consume. They will surely become addicted to WiwWiwWiw (What I want, when I want, where I want) media. The more they consume, the more they will pay. Some people will be OK with this; others will not.

As we know from experience, with regard to media and bandwidth, people with more money than time are willing to pay. However, people with more time than money are willing to steal. Piracy will run rampant. It should have been easy to thwart because most of the consumption devices will require monthly data plans. When a device is connected to a network and the network operator has your credit card number or billing address, the carrier could make it hard for you to steal content. In practice, the Stop Online Piracy Act (SOPA) and the Protect Intellectual Property Act (PIPA) have been soundly defeated by threats from the hactivist community. But some version of SOPA and PIPA will eventually pass, and it will cause an accelerated arms race.

The arms race will continue until the rigidity of the content-provider pricing and network-service-provider greed overwhelm consumers, or until a new technology evolves.

I don't think that mainstream consumers will ever value openness over certified convenience, but I do think that a new technology is going to evolve. I'm calling it the ProxyNet.

In this warped, "Shelly must be on drugs" vision of the future, connected devices are everywhere. The world is filled with billions of smartphones, app phones, tablets, PCs and wearable computers that come in only one flavor ... connected.

Sometime after 4G is deployed on a large scale, people are going to find a new kind of app available for every device. It will offer the ability to create a peer-to-peer (P2P) or Mesh connection to any device within radio range and not require (or even allow) an Internet connection. It will not use a commercial wireless network. It will not be 3G, 4G or any G. It will simply use the WiFi radio in the device to connect to other devices that are "friendly."

You can think of this network as a self-assembling, self-healing, invitation-only social media sharing network on steroids. It won't need the public Internet or a phone network, because it will simply connect to whatever device invites it to connect; hence the idea of a Proxy Network or ProxyNet. I really wanted to call it a "Subnet" because it's a network that exists under the radar. But "subnet" is a term of art in the network world, and it has a very specific and very different definition.

Anyway, the end of the world comes in the following form. People download the ProxyNet apps and start to form personal networks that almost mirror their social media networks. This happens automatically. If you want to think about it in a contemporary way, think about Foursquare or Twitter. If a ProxyNet app simply connected to your friends' devices when you were in proximity to them, you would not need the Internet or a phone company data network. The WiFi radios in the devices would do it all.

If this were the case, it would be easy to imagine five hundred students in a junior high school whose smartphones and laptops are all connected to each other but not connected to the Internet. The students would decide to share everything (1) 'cause they could and (2) 'cause it's free. You can't get into the network unless you're invited. You can't shut down the network, because it doesn't really exist. You can't get the identity of anyone on the network, because they are hidden by proxy (in the true network sense).

Now, imagine this on a global scale. Game over.

It may not happen exactly this way, but people are empowered by technology, and you can push them only so far. At a certain point, a few motivated 14-year-olds are going to build a browser plug-in, or a Facebook app, or just some little piece of code you can use to void the warranty on your phone or laptop and go digitally underground. I will be surprised if it doesn't happen. When it does ... well, I hope someone will figure out what we're all supposed to do after government work.

Traditional and New Media

"You're Too into Real Business to Understand Advertising and Social Media"

I had a very interesting discussion with a tech-savvy Internet veteran. He is a self-proclaimed "serial entrepreneur" with an impressive track record of successfully exited start-ups. For reasons of decorum and political correctness, I will call him Joe. It's not his real name (obviously), but this is a true story and you would recognize this person's real name instantly.

A mutual friend introduced me to Joe at a cocktail party. His new start-up is focused on reinventing advertising and marketing using social media. We were in violent agreement about the power and potential of social media, so I thought the conversation was going to evolve into a serious business conversation. I was wrong.

To my surprise and amazement (hyperbole intended), Joe presented his theory of the future of online advertising empowered by social media. He articulated a future where traditional media and big brands were dead, and just for good measure, he postulated a consumer-controlled, self-selected future of infinite choices of goods and services. His monologue was pervaded by phrases like "Advertising doesn't work" and "People don't want to be interrupted." And my favorite, "Social media marketing is the only possible future ... brands are completely out of control and marketers have to realize that consumers control everything! Besides, everyone wants targeted, measurable messaging; it's all going to be online."

Because of Joe's reputation and stature, I listened attentively for a while. Then I asked a question. "What do you mean by 'advertising'?" He answered, "You know, TV commercials and stuff like that." So I asked, "What do you mean by 'marketing'?" He answered, "Ideas to help sell stuff."

I don't think you can lump all advertising into one category, and I can assure you that marketing is more than just "ideas to help sell stuff." Definitions matter and semantics are important. And in this case, Joe simply didn't know what he didn't know. You may have a different way to describe the following (I hope you do), but here's my take:

First of all, advertising does work. Far from being one thing, there are at least three major types of advertising: brand/lifestyle, call to action and direct response (DR). The efficacy and return on investment (ROI) of DR advertising are accurately measured because you are asking the consumer to directly respond. So Joe and I did not need to discuss this $200+ billion use of advertising dollars. He admitted that DR is actually a pretty good business and actually works—but not until it was pointed out.

Then there's call-to-action advertising. This kind of message asks you to do something at a later time, such as "Come into our car dealership for a test drive and a free glass of soda this weekend and get zero percent financing." You can't directly respond, and it's not an emotional, ethereal, hard-to-measure, nontemporal request. You can measure the efficacy of a call to action by tracking how many people came into the car dealership before you ran your campaign, how many people came in during your campaign and what happened when the campaign was over. Call-to-action measurement is more of an art than a science, but it is measurable to a fair degree of accuracy. This really can't be what Joe was referring to, because this system is clunky and old-fashioned, but it's not really broken.

Perhaps Joe really meant brand/lifestyle advertising. It is hard to measure because it does not require a direct response, but there are plenty of organizations that will tell you how to measure the efficacy and ROI, and you know who they are. For example: How many BMW "Ultimate Driving Machine" ads do I need to see before I start dreaming about driving one? It's an important question, and it always will be. Perhaps this is what Joe really thought he was talking about. It's where $60+ billion gets spent annually. And to quote John Wanamaker ... well, you know the quote.

As my conversation continued with Joe, he informed me again and again that not only were big media dead, but big brands' days were numbered. "We are in the Information Age," he said; "consumers want to completely control what they consume."

While this is probably true for some percentage of items that we consume, let's not get crazy. You need to wash your clothes. You need to brush your teeth. You need inexpensive, nutritious food. You need durable goods, soft goods and hard goods, and you need them packaged and delivered to a venue near you ... every day. In fact, the more organic, unpreserved, special and perishable the produce and goods are, the more time and energy must be expended to get them as close to you as possible. Not just in your town, nationwide—and in some cases, worldwide.

So I asked Joe if he thought that people were giving up Cheerios or Tide or Crest or Ivory soap or cold cuts or spices or dog food or disposable diapers anytime soon. And then he said it: "You're too into real business to understand advertising and social media in the 21st century." You know, I never thought about it that way. "Too into real business?" I really didn't know what he meant by that, so I asked. He told me, in a very authoritative (and pejorative) way, that he wasn't really talking about the benefits of the Internet or social media for big, incumbent, multinational brands with just-in-time inventory systems and agile replenishment and RFID (Radio Frequency Identification) systems on loading docks. He'd never heard of "adjusted case volume" or "velocity at retail." ROI in Joe's mind was about the return on investment capital from the venture capitalist and private equity firms that substantially own his start-ups. He, to use his words, was talking about "all the new businesses that were forming in the Information Age."

I'm glad he cleared that up. As it turned out, he agreed that there is room for both of these types of businesses to exist. What a relief! We're not all doomed. Wow, that was close.

All kidding aside, this conversation clearly illustrates the profound difference between building a profitable business and building a business to profit from. Joe's market is the financial services community. The metrics he covets include numbers of unique users, frequency and loyalty. In Information Age Start-Up Land, you can fund and refund and exit a company with "big audience numbers" even if the company has no revenue or viable business plan to acquire revenue. Joe ascribes to the theory that something that is valuable to millions of people must really have a value, and you just need to get really big to figure out how to make a profit. While the Facebook IPO and Zynga and Groupon's fall from grace have thrown some cold water on the concept of "millions of users must have value," the idea lives on. And for many new businesses, it is still a popular model.

That said, no matter how much technology evolves, some businesses will be quick to adapt and others won't. Manufacturing, distribution, marketing, advertising and financing all respond to market pressures the best way they can. It's the nature of business. This doesn't matter to Joe; he's just trying to anticipate where the herd of potential investors is heading.

I think it is a huge mistake to focus or fixate on what the universe will eventually look like. Some of Joe's predictions about the future may be right, but his timing is not based on data—he's just guessing. It's not a good strategic exercise, and it's not a good practical or tactical exercise either. Most astrophysicists agree that at some point the sun will expand and consume the earth. Is it really helpful to put that in your long-range plan? So why try to guess when technological efficiency will usurp incumbent, massive, inertial systems? Why not simply profit from the delta between the two and use best practices at both? True digital leadership requires both knowledge of digital and the ability to lead!

I enjoy a Socratic debate about how technology is going to destroy everything we know as much as the next guy. And I get hired on a regular basis to scare people with stories of technologically empowered consumer behavior changes. But it's important to remember that the technology does not exist in a vacuum. As long as the VP of marketing at XYZ Company's main focus is to become the SVP of anything at Any Company, we are going to live in a world that doesn't work as designed … but works as evolved. So relax. You're NOT too into real business to understand social media. Most self-described social media types are too into technology to understand real business.

Understanding Traditional and New Media

Paid, Free, Owned and Earned

Traditional media have historically existed in two major categories: paid and free.

Paid advertising is airtime or ad space that you purchase. Free advertising is the editorial content that your public relations firm might have placed for you, and although you pay your PR firm, you generally don't pay for editorial placement.

New media include both types of traditional media (paid and free) but add two new categories: owned and earned.

Owned media are media created from the databases you own such as your website, email newsletters, apps, social media profiles, etc. Earned media are the media created by social media ambassadors of your business or brand. They are third parties (not generally paid professionals) that you empower to help get the message out.

Understanding Social Media

By now, you've probably noticed that the broadcast industry has fully embraced social media. Marketers are all over Twitter, as are celebrities and brands. Sadly, some people mischaracterize social media as just another channel of communication. They aren't. Social media allow us to engage in two relatively new forms of communication: many-to-one and many-to-many.

This is a nontrivial point. Most anthropologists believe that modern humans (a human being who would be physiologically capable of attending college today) have been on the planet for somewhere between 150,000 and 240,000 years. During that time we have been naturally selected to be expert at one-to-one communication. And although each of us performs according to our gifts, every healthy human being has the ability to carry on an understandable one-to-one interaction. We have evolved and adapted our physiology to be great at it.

About 3,500 years ago, the Greeks formalized the proscenium stage, and the modern construct of one-to-many communication came to be. I would argue that one of the highest forms of one-to-many communication is a broadcast television commercial. It has a beginning, middle and ending ... a rising action, climax and falling action, and it makes us buy stuff we don't need with money we don't have. Broadcast TV spots are awesome examples of one-to-many communication. And again, we are perfectly physiologically evolved to be great at it.

However, social media are not like one-to-one or one-to-many communication. Unlike one-to-one (conversation) or one-to-many (broadcasting) which humans are physiologically equipped to monitor, social media communication, which can be described as many-to-one (email, tweets, wall posts) and many-to-many (retweets, viral messages, etc.), requires digital tools to monitor. We don't have any physiology that would allow us to natively deal with social media. We must use email clients, Twitter clients, Facebook profiles and the like to interact with social media.

The Power of Social Media

"Information is not knowledge" is a recurring theme in my work, especially in the context of social media. For example, there is no point in tweeting "@yourname Joe Smith won the election." Everyone who cares about Joe Smith and the election will have access to that fact within seconds of the results being made public. Unless you are the local news authority, don't bother to tweet it. It is not adding any value for your followers. However, "@yourname Joe Smith won, so long after-school music programs" is an awesome tweet. It translates information (which is commoditized) into knowledge by adding editorial and context to the factoid.

To lead in a connected world, you must add value to your network. If you don't, you will find it exceedingly hard to succeed.

Speed and Scale

To fully appreciate the value of social media, you must appreciate the speed and scale of the networks that they can create. Simply stated, using social media, an individual now has the power to communicate on a global scale almost instantly.

Amplified Voices, Good and Bad

The speed and scale of messages on social media cause some people to refer to social media metaphorically as an amplifier. It's as good a description as any other. But it is important to remember that while social media amplify good ideas, they amplify bad ideas just as well and just as quickly.

Social Graph

Quite a bit has been written about the "social graph." It is supposed to be a representation of the way we interact through social media. Those who are proponents of the concept believe that analyzing the social graph of a population can reveal insights into purchasing preferences, behavior and other meaningful understanding. The goal is to monetize the assembled population.

Mark Zuckerberg, CEO of Facebook, has gone on record espousing the idea that Facebook will make money by monetizing its social graph. It's not working out, and the way Facebook is approaching it, it probably never will.

Trust Circles

One reason Mr. Zuckerberg is having so much trouble making money from social graphing Facebook users is that advertisers are not members of the trust circles that Facebook helps users form. A trust circle is the group of people you assemble around yourself for specific purposes. For example, in the offline world, you might know that Uncle Joey is the family car expert. When it's time for you to go car shopping, your mother will insist that you call Uncle Joey for advice. Online, you will assemble all of the "Uncle Joeys" in your social network and create a trust circle around the topic of buying your new car. Trust circles overlap, and you can have dozens of them operating seamlessly in an average social media account.

Some social media networks let you create special groups or subgroups for specific purposes. Google+ actually calls these groups "circles." No matter what they are called, for most people, there is no place for a brand or an advertiser inside them.

Traditional types of advertising get your attention by interrupting what you are doing. "We'll be back after this important commercial message." Creative advertising messages do a pretty good job in a traditional one-to-many environment, but they fail miserably when distributed on a many-to-one or many-to-many social network.

Virtual DMAs

One true bright spot in the new, omni-platform media landscape is the self-assembling of virtual DMAs.[1] A virtual DMA is a community of interest that self-assembles around a topic in social media. A virtual DMA does not have any geographic restrictions, so the "tyranny of geography" does not apply. This is a blessing

---

1 Designated Market Areas—a trademarked term of Nielsen Media Research, which divides the country into markets loosely based on population density.

and a curse. If you have a movie you want to stream, a virtual DMA allows you to reach all of the hand-raisers around the world at once. However, if you are running a sale for snow tire chains for a regional group of auto parts stores, you'd better make sure that the company has an awesome online and mobile customer experience, because a significant number of people in your virtual DMA may not be able to visit the retail locations.

Filter Success: The End of Social Discourse

While he was in Paris in March of 1789, Thomas Jefferson wrote a letter to Francis Hopkinson that scholars have aptly named "Neither Federalist nor Antifederalist." In it, Jefferson writes, "I am not a Federalist, because I never submitted the whole system of my opinions to the creed of any party of men whatever in religion, in philosophy, in politics, or in anything else where I was capable of thinking for myself. Such an addiction is the last degradation of a free and moral agent. If I could not go to heaven but with a party, I would not go there at all."

In the closing paragraph of this note, Jefferson asserts, "I never had an opinion in politics or religion which I was afraid to own. A costive reserve on these subjects might have procured me more esteem from some people, but less from myself."

I simply love the way Thomas Jefferson expressed himself. He was a man of many remarkable talents, a seriously independent thinker and, as I'm sure you can tell, a personal hero of mine.

Now, I don't like to discuss religion or politics in polite company. It's pointless. Both subjects provoke passionate lectures espousing personal worldviews, and minds are seldom, if ever, changed. But after watching some of the speeches at the 2012 Republican National Convention and switching between pundit commentary on CNN, Fox News and MSNBC, I started to wonder what @tjeff (not a real Twitter name) would have had to do to find the facts upon which to base his independent thinking. The red and blue filters are so good, and the mass and noise levels of the pridefully ignorant are so high, it is almost impossible to find a verified fact.

IMPORTANT: Before I start down this road, let me make one thing perfectly clear. I am not going to tell you which party I belong to, or which candidates I support, or make any suggestions about whom you should vote for. This is not a political or religious writing. It's just about information filters and fact-finding in the 21st century.

Like many of you, I believe that elections are important. And although conventions are not as interesting as they used to be, I was excited to see how the Republican Party was going to present itself. Putting on my independent thinker @tjeff hat, I wanted to watch the least-filtered, least-punditized broadcast of the convention.

On Day 2 (August 29, 2012), the major broadcast networks were not going to cover the convention until 10 p.m. EST. So in order to watch former Secretary of State Condoleezza Rice speak, I turned on CNN around 9:30 p.m., hoping for nonpartisan color and commentary. CNN is still a filter, but I was too lazy to go searching for CSPAN. I suffered through CNN's coverage of Ms. Rice's speech. I say "suffered through" because CNN had a lower-third graphic with all kinds of editorial nonsense flashing on the screen throughout the entire speech. I suppose that flashing select résumé items for Ms. Rice under her speech was helpful to some viewers, but from my point of view, it was distracting, detracting, gratuitous noise that was insulting to educated viewers and disrespectful to the speaker. When she finished, I switched to NBC to get a full HD picture with no editorial graphics.

Next up was vice presidential hopeful Paul Ryan. During Representative Ryan's speech, he told a story about candidate Barack Obama four years ago. Here's the quote:

"My home state voted for President Obama. When he talked about change, many people liked the sound of it, especially in Janesville, where we were about to lose a major factory. A lot of guys I went to high school with worked at that GM plant. Right there at that plant, candidate Obama said: "I believe that if our government is there to support you… this plant will be here for another hundred years." That's what he said in 2008. Well, as it turned out, that plant didn't last another year. It is locked up and empty to this day."

According to the *Washington Post*, the decision to close the plant was made in June 2008, when George W. Bush was president. Ryan says that Janesville was "about to" lose the factory at the time of the election, and Obama failed to prevent this. This is false, as Ryan knew in 2008 when he issued a statement bemoaning the plant's impending closing. Republican fact checkers say that the Washington Post fact checker is wrong and that the plant did close in 2009 on President Obama's watch. What would an impartial viewer of the speech have thought? What would you have thought?

Now, just for a second, put your politics aside. Forget whose side you're on and think about this the way @tjeff would.

- **MSNBC** vilified Ryan—not unexpected; it is the blue state channel.

- **Fox News** reported it, but dismissed it as a mischaracterization—not unexpected; it is the red state channel.

- **The Interweb** (Twitter, Facebook, blogs, etc.) split precisely along party lines.

None of this is surprising, nor is any of it news. Politicians lie all the time no matter what party they represent. But with all of your filters in place, how can you find the truth?

Download the transcript of Representative Ryan's speech and fact check it for yourself. I did, just like @tjeff would have, and it took hours. It would have been so much easier to just let my preferred filter do the work for me. It would have been blissful to blanket myself in the comfort of the information that I wanted to hear. It would have been super easy to cocoon myself inside my filter set. If I just lay back and enjoyed it, I too could be loud, proud and pridefully ignorant (if I were ever to let myself discuss politics in public).

But just this once, just to confirm my own thesis that filter success is speeding the end of social discourse, I spent the required time with the transcript of Representative Ryan's speech and my computer. After two hours of searching the "facts" in the representative's speech, I can say with authority that it will be hard for you to find verifiable facts with two sources that would allow you to take this speech at face value (and I am being kind).

Again, this is not a political comment on any particular thing that was said, only a comment on how it was filtered.

Here's the really unfortunate part. It is abundantly obvious that the vast majority of people don't care. Why bother to fact check? Why take the time to form your own opinion? You can let Rachel Maddow (blue states) or Bill O'Reilly (red states) do it for you. If you spend any time reading the millions of words written about this single speech (and because everything is online, it doesn't matter what year you read this book, it will all be easy to find), you will quickly realize that politics and religion share a single attribute—they are faith based.

What is the truth? Whose truth? With whom can you debate and discuss the issues? Where are the facts? Is civil, social discourse the victim of social media? And if it is, are there tools and techniques that digital leaders can use to save it? The filters we have evolved are so personal and so good; "Such an addiction is the last degradation of a free and moral agent."

Pandering to a "least common denominator" on a global scale is a frightening thought, but there is a ray of hope ahead. We, as a society, are digitally maturing, and some of us are learning to use social media in a more mature way.

Personal Facebook Cleanse: The Evolution of Social Media

I was thinking about a juice cleanse. Everyone is doing them. They are all the rage. Of course, my doctor talked me out of it. He told me that my liver is healthy and minding it is his job. He went on to emphatically state that there is no truth to the marketing mythology about detoxification through juice. Bummer ... I thought it sounded awesome.

While juice cleansing was on my mind, I tapped my social networks to learn about cleanse success and failure. The results were mixed and helpful, but during the process, something else occurred to me: It's time for a Facebook Cleanse.

What's a Facebook Cleanse? Well, there are several different protocols, but the one I'm thinking about is simple: It involves posting a message to my personal profile saying something like, "I'm doing a Facebook Cleanse, like or comment here if you want to remain my Facebook friend." If people don't like the post or comment, I am going to unfriend them. The goal is to achieve fewer than 250 Facebook friends. Why?

Well, I think there's a reason that I fell out of touch with most of my grade school and middle school friends. I'm also not sure that the PTA people I friended when my kids were in school still need to be in my inner circle. Some have become lifelong real friends, but the others were really only there so we could sort out who was handling snack after games and meets.

After I go through this exercise, I'm going to spend some quality time with my news feed and see what items interest me. If I don't want to comment or like after three posts from the same person, I'm going to unfriend them.

The result should be a very small group of people whom I really care about and who really care about me. These should be people whom I know well—well enough to hang out with whenever they're in town, well enough to want to share family gossip with. My Facebook Cleanse is going to be awesome!

But what will happen if I want to find and refriend one of my semi-friends or acquaintances? What if I delete someone I really should stay in touch with, even though I don't? What about the backlash from the group that likes getting my updates, but doesn't really interact much? What about people who will simply be offended that I unfriended them?

I've never done a juice cleanse, but I can imagine that the process is not hugely pleasant. I imagine that some bodily functions are dramatically impacted by ingesting only juice for five days. My guess is that a Facebook Cleanse will be equally uncomfortable. But unlike a juice cleanse, a Facebook Cleanse will yield real benefits.

First and foremost, I will remove most of the noise from my Facebook newsfeed. This will allow me to assess the state of my social network at a glance. It will also make my mobile experience much, much better. Then, because my network is now made up of trust circles that I influence and that influence me, I will be able to quickly learn if any given message adds value to the network or annoys it. This learning is invaluable. Finally, I will be able to leverage the people I influence more directly with better-crafted messaging and better harness the power of my social media network.

As for a down side, I may ruffle a few feathers, but it won't be fatal. Digital leadership means making choices about how to most efficiently use your digital tools. This is a good training exercise.

A Social Network Is Not a Popularity Contest

A close friend of mine owns a high-end print shop, one of the few left in the world. He specializes in printed invitations for weddings, sweet sixteen parties, etc. When AIM was in its heyday, the average number of AIM buddies was 125. Back then, he told me that the buddy list number was virtually identical to the average number of people who would receive printed invitations to an important life event. I always thought that was an interesting statistic.

But that was then. How about now? According to some recent numbers prepared by Pew Research Center's Internet and American Life Project, the average number of Facebook friends in various generations goes something like this:

• 318.5 – Average number of friends a Facebook user who is a Millennial or Gen Y member has (ages 18–34)

• 197.6 – Average number of friends a Facebook user who is a Gen X member has (age 35–46)

• 124.2 – Average number of friends a Facebook user who is a Baby Boomer has (age 47–65)

• 78.4 – Average number of friends a Facebook user who is a Silent Generation member has (age 66–74)

• 42.0 – Average number of friends a Facebook user who is G.I. Generation member has (age 75+)

How remarkable that for Baby Boomers, the number of Facebook friends looks the same as the AIM buddy number and the average number of invitations to important life cycle events. What's the right number? What's your right number? Is it time for your Facebook Cleanse? Of course it is ... and I hear it goes great with juice!

Flames and Techno-Politics

Tweetmobs: The Attack of the Fifth Estate

Can any traditionally organized government withstand the ideological forces of a technologically empowered proletariat? Call them tweetmobs, blogmobs, txtmobs or simply citizen publishers—everyone has a voice, and no matter how the government tries to silence them, in the Information Age, information will find a way.

Throughout history, the Estates of the Realm have come under various attacks. In the past, the asymmetry of wealth usually struck the decisive winning blow for the First or Second estate, and all was set right with the world. Truth be told, the Fourth Estate (the press) has been under the complete control of the First (clergy) and Second (nobility) Estates throughout most of history, so the control of information and the control of knowledge were pretty well guarded and defensible.

It is our freedoms, such as freedom of religion and freedom of the press, that make America such a grand political experiment. To exercise and express our freedoms, we have been using technology, which heretofore almost exclusively emanated from a central point. Be it a leader or an associated bureaucracy, there was almost always a yellow brick road leading to a physical edifice. This is no longer the case. Information is the true currency of our age, and it has transcended its bounds.

Far from being in the control of the realm, information now displays many of the characteristics we attribute to living things. It is born, it evolves, it eats, it excretes, it mutates—but, interestingly, it can no longer die. This was true on a small scale even before the "great unwashed" learned to blog. However, this is the first time in history that an idea (good or bad, true or false) can travel worldwide instantaneously and live on with a permanent, un-erasable record of itself. Burn all the books you like; the knowledge lives on in the ether.

One of the underlying principles of a democracy is that the majority rules, but the rights of the minority are always protected by assuring them the right to speak and to vote. This principle is closely tied to another convention of democracy that requires all citizens to abide by the majority rule, even if they did not vote with the majority. We have the right to disagree with our leadership, and our laws give us a way to show them our displeasure with their performance—we can vote them out of office. Truly, one of the most remarkable and enviable attributes of our democracy is the peaceful transition of power.

Through wars, depression, recession, boom and bust, America has survived and ultimately thrived. But it has never had to deal with a decentralized, self-organizing, cogent, antagonistic Fifth Estate until now. Tweetmobs or blogmobs form like regular mobs. They can be constructive or destructive, patriots or rebels. But unlike their flesh-and-blood counterparts, their ideas can have immense, virtually instantaneous impact on a worldwide platform.

The most interesting attribute of tweetmobs is the way they self-assemble. These swarms of ideas evolve into points of view and then metamorphose and mutate into the next phase of their existence—gaining or losing the power to influence their audience. It is a fascinating twist on our ability to communicate, and for organizations that rely on central control (like governments and corporations), it is not necessarily a good thing.

The tweetisphere and the blogosphere are replete with compelling arguments for and against everything—a true adhocracy—no gatekeepers, no pundits, no leadership, just passionate points of view. Perhaps the legacy of microblogging sites like Twitter and RSS (Really Simple Syndication, the specification that enables blogging) will be the empowerment of a true Fifth Estate with a collective mind of its own and the ability to amplify the voice of the people above all others. This raises the question: "Can any traditionally organized government withstand an ideological force of this magnitude?"

The Dark Side

Pornbot Twam: Twitter's Dark Side

Elfrieda Utley @lerahxg followed me. She's from Arkansas. If you don't look at her Twitter profile too carefully, it looks pretty normal. She has 128 followers, she's following 1,407 people and she has tweeted 43 times. Her last tweet, "I can remove 90 percent of your so-called beauty with a Kleenex," is stupid, but sadly, stupidity is not a crime on Twitter. That said, if you do look carefully at her Twitter profile, there is a problem, and it's pretty significant: @lerahxg is a pornbot. Its website is http://best-xxx-vids1.info. Ugh!

By the time you read this, the account will probably be suspended—but maybe not. There's a slight chance that Twitter won't find anything wrong with it (other than the pornographic content of the website).

Elfrieda represents so many problems with Twitter I don't quite know where to start.

First and foremost, the content is available to anyone of any age. Twitter should not be linking underage people to porn. But there's more. @lerahxg is a bot, so it falsely inflates follower metrics. Bots are counted, but they don't count. Don't you wonder how many pornbots are following your favorite celebs or politicians? I promise you, it is a very big number.

I probably identify 50 new followers as pornbots every week. Twitter gives you two options for dealing with this kind of annoyance. You can "Block @name" or "Report @name for spam." At this writing, if you report the offending follower as spam, the follower is automatically blocked and removed from your followers.

The sad news is that you really have to look at every single person who follows you and determine if that person is a real person or a bot.

Link Twammers

Of course, not all bots are pornbots. There are plenty of other types of twam (Twitter spam). There are "link twammers" who use your @name to get your followers to click on their link. These really make me mad because they are trading on my brand name. When you see a tweet like @shellypalmer thanks so much, check this out http://ow.ly/b30eA, you should check to make sure that you've interacted with the person. If you haven't, don't click on the link; just report the tweet as spam because there's a very good chance that the link is malware or spam and that you are being link twammed.

@reply Bots and Trending Tweet Bots

Then there are @reply bots and trending tweet bots that twam you or your followers as the names imply. @reply bots are far more offensive than trending tweet bots because @reply bots actually reply to your tweets.

Other than reporting them to Twitter, there is not much you can do to protect yourself. But protect yourself you must! No one else can do it for you.

If you use Twitter for business, it is essential that you remove all offending twammers from your followers as soon as you identify them. Why?

Twitter is both a medium and a metric. You use Twitter to communicate, but you also use Twitter to measure the efficacy of your social media. A bogus follower count (especially one that is inflated by pornbots) does not accurately represent the value of your social media communication. And should a quantitative entity analyze your Twitter account, you will look very foolish to your boss. Pornbots can get you fired.

We analyzed 25 random Twitter profiles, each with more than 20,000 followers, and found that all of them had a substantial percentage of bot and pornbot followers. One account had over 35 percent bot followers, which is pretty sad. If you factor in other types of bot followers, you can quickly get to a metric that is 50 percent lower than your "official" follower count.

If you project these numbers out to the greater Twitter community, it doesn't take long to come to the conclusion that the number of followers someone has is not a meaningful number, unless it is adjusted for bots and twam.

Do yourself a favor. Grab a tool like tweepi[2] and analyze your Twitter account. When you're done, I'd love to hear from you. How pervasive is twam in your neck of the tweetisphere? LMK @shellypalmer.

Digital Parenting

Sexting Is More than Pix

Can you translate this dialog: kotl. iwsn. gypo. l8r. now. 2 c-p. 459. ruh. 143. im so fah, gypo. lmirl. no, gnoc. pir. ttfn? (Answer key at the bottom of the section.)

Don't you speak Sext? Most of the 13- to 19-year-olds in America do. Add a still picture or video taken in the shower and you have all the ingredients you need to publish what used to be a very private moment.

As you well know, technology is seamlessly woven into the fabric of the lives of every "Born Digital," kids born after 1989. You can no longer ask, "Should we get little

Johnny a cell phone?" The question now is, "Which cell phone is best for little Johnny?" You can't protect or shield kids from technology—it is a pervasive force in our culture. And perhaps most importantly, you cannot alter how people's behavior will evolve with technology. You can only seek to understand it and make both the upside and downside known.

I'm sure some of you will push back on the last point. After all, in our society we use the rule of law to regulate harmful acts and even harmful tools. You need a license to drive a car, buy a gun (in most states); you need a prescription to legally purchase drugs, and you must have attained the age of majority to purchase tobacco products and alcohol. We even have child pornography laws to protect our children from sexual predators. I'm stating the obvious, since most of us are completely aware of the laws surrounding the doing of life. You shouldn't drive drunk. You shouldn't break the speed limit. You shouldn't download music or movie files you don't have rights to, etc.

The problem is that police rarely come to your home and arrest you for file-sharing and you have to get caught to get a speeding ticket. When a 15-year-old girl sends a video of herself, naked and doing seductive things, to an 18-year-old boy she hardly knows, what should happen? She has broken any number of child pornography laws. So has he. What to do?

Moving on, if you are paying for your teen's cell phone, should you have the right to read (and decode) the opening paragraph of this article? What would you do with the information? Would you listen to that conversation if it were a voice call? Would you eavesdrop if the conversation took place on your living room couch?

There was a "sexual revolution" in the '60s. It was a decade of transition. Hippies transformed into disco queens, LSD went out of vogue, cocaine became the coin of the realm, and each subsequent decade had its own kind of revolution. In the 21st century teens are empowered with media tools—and they are using them in extremely social ways.

Before you can find a solution, you need to identify that you (all of us) have a problem. It's a simple one best described by one of my favorite George Bernard Shaw quotes: "Every profession is a conspiracy against the laity." In this case, teens are the social media professionals and we are the lay public. Do you speak 14-year-old? Perhaps it's time to learn.

kotl: Kiss on the lips.

iwsn: I want sex now.

gypo: Get your pants off.

l8r: Later.

now: Now.

2 c-p: Too sleepy.

459: I love you.

ruh: Are you horny?

143: I love you.

im so fah: I'm so f***ing hot.

gypo: Get your pants off.

lmirl: Let's meet in real life.

no, gnoc: No, get naked on camera.

pir: Parents in room.

ttfn: Ta ta for now.

The Nontechnical Solution to Sexting

Sexting is slang for sending and receiving sexual content using mobile phones. How do you do it? It's very simple, really. You either send txt messages containing untoward content, or play a modern game of I'll show you mine if you show me yours, with the camera in your mobile device or computer. For a while, sexting was a problem for kids. Now it is a problem in everyone's connected life, and I have some suggestions.

First, I think it's important to understand the technology. This may make you roll your eyes. But do you think Tiger Woods really understood that the voice mail he left (that we've all heard) was a digital recording of his voice? That it was recorded on someone else's device and that he no longer had control of it?

A little history: As you know, Thomas Edison invented the phonograph back in 1877. What you may not know is that the early phonographs could both record and play back (very much like a modern-day voicemail system). Edison liked to demonstrate his phonograph by allowing people to speak into the machine and then playing the recording back for them. However, the technology was outside almost everyone's conceptual understanding. Up to that point in history, the only known method of creating a disembodied voice required a ventriloquist, so people thought it had to be a trick. Clergymen came to pronounce it "the devil's work" and to discredit Edison.

But here's the really fun part: Edison used to charge people 25 cents to try to "fool the machine." A person who spoke Latin (a dead language) would speak Latin into it, and of course it would speak Latin back to the person. People wondered how Edison was able to teach a machine to speak Latin. A person would speak Chinese into the machine, and it would speak Chinese back to them. Again, people would wonder how the "Wizard of Menlo Park" taught the machine to speak Chinese. People simply did not understand the concept of a recording.

That was then. Now it's the 21st century. What was Tiger thinking? Didn't he understand that he was making a recording? Understanding the technology may help you think about how you want to start the conversation with your kids.

All of the mobile devices we're talking about are digital. They are little computers. When you send a txt message, you can think of it as a word processing document that is automatically stored and delivered. The only problem is, you don't know where it is stored and to whom it will ultimately be delivered. The same goes for digital pictures. Cameraphone or digital camera, the image is uploadable and downloadable. And once you press "send," you can never get it back. No matter what you do, no matter whom you know, and now matter how hard you try. These are computer files, and they are as easy to share as music and video and pictures you upload to Facebook.

I saw a very effective demonstration that the parents in one family did for their 16-year-old daughter. At her sweet 16 party, the obligatory photomontagey included a picture of her in the bathtub at age 6 months. She was naked, of course. Her friends giggled. She was mortified. The next day, her father asked her if there were any other pictures that would embarrass her if they were displayed at the party. She blushed and said she didn't think so, then grabbed her phone and ran into her room. Hopefully she remembered to empty the trash on her phone and her computer as well. As you know, just deleting something does not remove it from your device.

There is no technological solution to this problem. But there is a very reasonable parenting solution. When my kids were little and it was time to cross the street, I asked them to hold my hand as we crossed. We've all done this with our kids. But that wasn't the only thing we did. We instructed them to look both ways. We discussed the consequences of crossing a busy street without looking. We didn't do it once; we did it almost every time we crossed the street with them. And we did it for years. As they got older, there were some streets we could cross together without holding hands, but even with a certain degree of autonomy, if we came to a very busy street, hands were held.

One day, one very important day, they were allowed to cross the street by themselves. This day did not just happen. The trust was earned over a protracted training period under our very watchful eyes.

We don't want to stop anyone from using and exploring new technology. But kids should not be allowed to "play" with power tools, and that's what these devices are. They are digital tools that make permanent records of how we use them. As parents and school administrators, it's up to you to make this simple fact sink in. You can't have one short chat about it; you need to hold their "digital" hands and make sure that they understand the danger of publishing content about themselves and others that can never be unpublished.

In the 21st century, nothing can be unsaid.

Social Media Policies Should Be Guidelines, Not Rules

One of the biggest issues facing businesses today is how to deal with employees and customers in a connected world. Almost everyone has access to social media. This is not a fad; it's a trend that will continue until literally every one of us is connected.

This level of instant, public connectivity comes with a significant set of unique problems. First and foremost is: "When are you an employee of your company, and when are you just a person?" In the old days (five years ago), you could go to a bar after work and say something to a co-worker, a colleague, a friend or even a stranger and not think too much about it. Today, there is a very good chance that anything you do will be photographed, video recorded, audio recorded or simply tweeted about.

The most common variation of this issue is having an employee simply respond to a tweet, blog or social media post without a proper briefing on corporate policy. As we all know, in the 21st century, nothing can be unsaid or unpublished.

The list of bad things that can happen when people have access to powerful publishing tools is endless, but there is a simple way to deal with it.

While rules "are made to be broken," guidelines are easy to set and easy to follow. Corporate governance around social media is easy. Ask every employee to behave as if every post is crafted by their job title and not by them personally. Or ask everyone to behave as if they are already famous and every word they write and everything they say will be as interesting to the media as a Kim Kardashian sex tape, and everything will take care of itself.

Succeeding in a connected world requires thoughtful leadership through new, uncharted digital territory. Corporate governance may not yet have evolved or adapted to the need for social media guidelines, but adapt it must. Will social be owned by the public relations department, advertising, marketing, sales, customer service? In theory, social should be owned by everyone, but in practice there are fiefdoms and competitive stakeholders. Want to impress your boss? Figure out how to appropriately govern social media throughout your organization. It's an awesome exercise that will yield real benefits.

Of course, before fighting for appropriate governance, it might be useful to understand the value of social media to you and your company. Do social media have a value? Let's ask Kim.

The Kardashian Effect: A Social Media Conundrum

What is the value of a Facebook "Friend"? What is the value of a Facebook "Like"? What is the value of a Twitter "Follower"? How do you calculate the ROI? What is the value of a fan engagement? How do you quantify time spent on your site, with your profile or on your fan page?

These questions are "Topic A" at almost every meeting I attend these days. Everyone knows you need to be doing something in social media, but it is increasingly hard to quantify its value.

If you "like" me on Facebook or follow me on Twitter, you will be one of several thousand fans who interact with me on a daily basis. On any given day, I post stories and links to things I find interesting. And on any given day, a handful of people will post a comment or just press "Like" on a post that resonates with them.

This social interaction is awesome. It keeps me in touch with my most loyal brand ambassadors. It gives me instant feedback (both positive and negative) about the things I'm doing. And most importantly, it gives me a way to learn about and interact with the people who are interested in the things that interest me. It's win/win.

But wait! I'm supposed to be a social media "expert." Social media are a huge focus of my consulting practice. How come I don't have a zillion fans and two zillion followers? What's wrong with me?

Actually ... nothing. I'm doing just fine with the fans ("likes") and followers I have. My fans and followers are growing organically each day, and it's a great group of people.

That said, if I wanted a zillion fans and two zillion followers, I could obtain them in about a week. But I don't really want them and neither do you.

To prove this theory, I decided to test the Kardashian Effect. It's a term I use to describe the throngs of useless fans and followers that can be obtained by confusing the "famous for being famous" with actual people. Anyone who wants a zillion fans and two zillion followers need only look at the trending topics and create a well-SEO'd, well-SEM'd social and online presence focusing purely on them. The results are instant and obvious. You will get a ton of traffic. There's only one problem—you won't be able to keep it unless you do it every day forever. If you're not offering them the best place to get a full dose of the thing they crave (which is the hottest thing available), they won't stay and they won't be back.

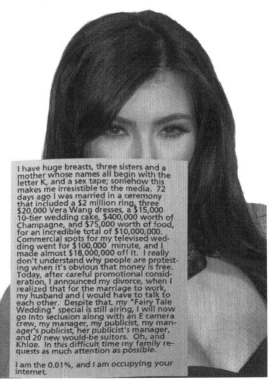

I have huge breasts, three sisters and a mother whose names all begin with the letter K, and a sex tape; somehow this makes me irresistible to the media. 72 days ago I was married in a ceremony that included a $2 million ring, three $20,000 Vera Wang dresses, a $15,000 10-tier wedding cake, $400,000 worth of Champagne, and $75,000 worth of food, for an incredible total of $10,000,000. Commercial spots for my televised wedding went for $100,000 minute, and I made almost $18,000,000 off it. I really don't understand why people are protesting when it's obvious that money is free. Today, after careful promotional consideration, I announced my divorce, when I realized that for the marriage to work, my husband and I would have to talk to each other. Despite that, my "Fairy Tale Wedding" special is still airing, I will now go into seclusion along with an E camera crew, my manager, my publicist, my manager's publicist, her publicist's manager, and 20 new would-be suitors. Oh, and Khloe. In this difficult time my family requests as much attention as possible.

I am the 0.01%, and I am occupying your Internet.

On top of that, you must commit to finding ways to translate the value of transient Kardashianites into wealth. (Your wealth, not hers.) Think this is easy? Think again. It's very easy to create value online, but creating wealth (especially your wealth) is exceptionally hard.

Anyway, the test was simple. A friend of mine sent me this picture of Kim. It is obviously "fun with Photoshop." It is credited to twitter.com/kelkulus from Los Angeles, CA. His description says: "Just in case this comedian thing doesn't work out, I'm also studying to be a rockstar." I don't know @kelkulus, but I thought the image was pretty funny, so to prove the Kardashian Effect is real, I posted it on my Facebook wall and tweeted it as well.

As suspected, the results were instant and obvious. The highest number of likes, highest number of comments and highest engagement metrics of anything I had posted to date. People love to hate her. People think she's a scam artist. People think she's a genius. The nature of the responses is not important. What is important is that people responded—like crazy.

What does that mean for my business (or yours)? If I want tons of transient traffic and useless interactions, I can post more stuff about Kim or Lindsay Lohan or (substitute your tabloid star here) and put huge but ultimately useless metrics on the scoreboard. Or I can just do what I do and continue to interact and profit from my loyal fans and followers.

The lure of the Kardashian Effect is overwhelming. I really want zillions of fans and two zillions of followers, but I also want to be respected for my work, keep my brand message clear and understandable and profit from my marketing efforts. What to do? What to do?

I know. I'll trick my advertisers with these inflated numbers and make them think that my social media efforts are stellar. Oh, wait, I don't sell any advertising ... hmmm.

All kidding aside, the Kardashian Effect is real. Kim is a genius and we are all drinking the social media metrics Kool-Aid. Qualitative understanding of social interactions is key to unlocking their value. The goal is the target, not raw tonnage. And most importantly, quality organic social touch points should never be sacrificed for scale. You can never sustainably profit from that kind of Pyrrhic victory.

Measuring Social Media Tools and Techniques

It's not easy to measure the efficacy of social media. Measuring ROI is even harder. Conventional wisdom says that ROI for social media is best measured when social media are used as a customer service tool. Their value as a marketing tool is still questionable. That said, there are some best practice ideas and tools that should be on your radar if they are not already fully implemented in your organization.

Listening Posts

Every business should have a set of social media listening tools deployed and specially configured for each department in their organization. Executives should do this for themselves as well. A social listening post can be as simple as a free tool like Tweetdeck, or you can find several low-cost commercial listening tools and services. The theory is simple: Listen. There is a massive global conversation going on 24/7, and you need a listening post to hear it, so set some up.

Measurement

There is a school of marketing that says if it can't be measured, it doesn't exist. I don't ascribe to this philosophy. For those who do, there are literally hundreds of companies that will offer ways to measure the value of social media. Some of these services legitimately try to further the art and science of marketing. Most are snake oil salesmen.

In almost all cases, you can measure activity. The number of tweets is sometimes referred to as "mass," and the frequency is often referred to as "velocity." This is all jargon, and whether you describe your social media tracking using terms like "volume" or "mass" or "number of tweets" really does not matter. What does matter is sentiment (good or bad), the number of tweets as a ratio to total tweets in the period and duration of the "news cycle." One interesting thing about social media is that while vicious, evil, negative waves of posts known as Flames can appear out of nowhere and become enormous almost instantly, they often disperse just as quickly.

I am not suggesting that you can ignore a tweet like "Toyotas kill people!" You can't. But you have to be careful not to fan Flames and not to knee-jerk react to real-time information.

Engagement

"Engagement" should be banned from marketing nomenclature. It is a meaningless, catch-all word that represents a concept. It is not a metric. Because it is not a metric, there is no way to charge an advertiser for it. "The audience was engaged with our content." Really? What does that mean? "Our goal is to increase audience engagement with our content." I see. What does that mean?

Reduced to practice, let's say you had 20 million votes in the first episode of your reality show and 30 million votes in the next. How would you charge for the extra "engagement"? Certainly the audience was more engaged in the episode in which 30 million of them voted. But if the ratings did not change, how would you convert the concept of engagement into cash? Engagement is a concept, and the definition is specific to each person who uses the word. Let's get an industry index we can use to monetize engagement or stop using the term in the context of media sales.

Targeting

One of the outstanding traits of data-driven marketing is the ability to target very specific audiences. This is done with a variety of tools. If you want to scrape social media feeds, you can very quickly figure out what people are interested in and exploit that knowledge. Some marketers believe that highly targeted listening tools yield excellent opportunities to target and retarget.[2] Other marketers believe that identifying self-assembled trust circles or communities of interest and earning your way in yield better results. Only time (and privacy laws) will tell us which school of thought is right. My suspicion is that consumer control of incoming messaging is now so enhanced that these decisions are probably best left to brand managers. In fact, for best results, most decisions about targeting should be made at the SKU[3] level.

Return on Investment (ROI)

Measuring ROI is a topic that comes up in almost every meeting about social media. If you use social media for customer service, it's pretty easy to measure how much time and what kind of resources social media are adding to or replacing. On the other hand, when looking at social media as a public relations or marketing tool, the calculations become significantly more difficult. Again, social media are not just another channel; they are a completely new form of communication. Smart managers and marketers are jumping deep into social media and

---

2 A form of online advertising also known as search retargeting, behavioral retargeting or simply retargeting that serves online ads to users who have exhibited specific behaviors or searches or visited specific sites.
3 SKU is an abbreviation for Stock Keeping Unit. It is a unique identifier for each product or service sold.

learning all they can learn about how they work for their businesses. There is no single right way to do social media. There is no single right way to measure ROI. It is unique to each company culture and brand, and as in targeting, may be unique to single SKUs.

Tweet vs. Retweet

Social media pundits don't agree on a lot of things, but one area of consensus is that retweets[4] are more valuable than tweets. This makes intuitive sense. If you've tweeted something of value, the best way to measure that value is to count the number of people who also think it is valuable. It's pretty simple: The more tweets that are retweeted, the more valuable you are to your followers.

Social: Medium and Metric

Social media have two functions in our world today: They are both a medium and a metric. You use social media to publish your messages, but you also look to social media metrics as a measure of success: How many friends or fans or likes or followers, etc.?

This is fine in principle, but the power of a social network comes from its quality, not its size. A reasonably sized, high-quality social network of brand ambassadors will add immense value, while a gigantic, superficially interested network cannot be activated and is therefore far less valuable.

---

4 On Twitter, the act of placing an RT in front of someone else's tweet and sending it to your followers.

Customer Service in a Connected World

Hertz: An Unqualified Success in Customer Service

A while back, cruising the highways of Southern California in the Ford Mustang GT500 I rented from Hertz, I was forced to skip two unmanned tollbooths. As crimes and misdemeanors go, this loathsome act probably would not have landed me on the "most wanted tollbooth skipper" list. But I hated the fact that I owed California $2.50 ($1.25 for each toll), and I wanted to do the right thing. I also knew that both tollbooths took a picture of me and my license plate, so there was no way I was going to get away with my heinous crimes.

When I finally exited the Toll Roads system, I received a card with the URL http://thetollroads.com/missedtoll. The solution should have been simple and painless. It wasn't. It was so painful, in fact, that it prompted one of the most popular blog posts I've ever written, titled TheTollRoads.com: An Abject Failure in Customer Service..

I am always astonished at how quickly bad news and stories about bad experiences travel. The story was picked up and shared by hundreds of people. It was well amplified by social media as well as traditional media. The total reach wasn't big by current standards, but it did strike a chord with many people and did reach beyond my usual network.

This is not news, nor is it particularly interesting, but here's the interesting part: Within a week of my blog post, I received a letter via snail mail from Allen P. Thomas, Sr., Supervisor, Hertz Customer Service that is so extraordinary I want to share it with you.

Dear Mr. Palmer,

We recently saw your post on the Huff Report dated June 29, 2012, and here at Hertz we always want to help our customers when they're in our cars, even if it isn't exactly our issue. I was sorry to see you had such a problem with paying your tolls.

I have enclosed $50.00 in discount coupons to apply on your next rental with

Hertz. I hope your next visit to California or Hertz rental destination is free from these types of distractions.

Thank you for using Hertz for your rental car needs.

Allen P. Thomas
Sr. Supervisor
Hertz Customer Service

In my original story, I did mention that, because I was in a rental car, the license plate had too many violations associated with it to allow me to pay online. But I did not blame Hertz, nor mention them in anything but the best light. I am a Hertz Gold Club member, nothing special; in fact, the only benefit I receive from Hertz is the convenience of walking directly to my car. I'm loyal to Hertz, but only because of the professional way it treats its business customers. I don't have a special deal or special rates—I'm just an ordinary low-expectation, low-profit client for them. I mentioned them in my blog post because I was really enjoying the Ford Mustang GT500 they hooked me up with, so the car (and Hertz) were on my mind. And no, Hertz did not upgrade me for free—I saw it on the lot at the Long Beach Airport, was excited that they had the GT500 available (who can resist an overpowered Mustang while driving in CA?), and paid for the upgrade.

If I was a loyal Hertz customer before this episode, you can imagine how I feel about Hertz now. I don't know Mr. Thomas, and I might never communicate with him, but Hertz literally struck gold. For setting up a listening post, paying attention to the Interweb, the tweetisphere and Facebookistan plus $50 in coupons, Hertz now has a true brand ambassador and a vocal advocate with a story to tell.

How simple was it for Hertz to accomplish this? Very. You can do it for your business with Google Alerts for free or with any of a hundred paid tools. Hertz could also have simply used TweetDeck or SocialOomph or Hootsuite, or any of a hundred free or low-priced social media listening tools. These programs are so easy to set up and monitor, there is absolutely no excuse not to use them.

If you're interested in building brand ambassadors for your business, learn to set up social listening posts and use them. It is truly a requirement for success in a connected world. Wondering about the ROI? Just ask Hertz or me. Remember, if you don't take care of your customers, someone else will.

Cloud Life Chronicles Part 1: Hijacked and Killed on Facebook

What happens when you drink the social media Kool-Aid? OK, better question: What happens when social media become woven into the fabric of your life and you rely on a cloud-based profile (that you don't think to back up or export) as the center of your social universe? Answer 1: When everything is working as designed, the world is virtually wonderful. Answer 2: When everything is NOT working as designed, we get an up-close, personal view of the pain and anguish a truly online, cloud-based world could eventually offer us. As bright as the upside is, the dark side is darker!

I've been on Facebook for as long as they have allowed grown-ups. (You can "like" me at facebook.com/shellypalmerdigitalliving.) Like most people who lead connected lives, everyone in my family, in fact, almost everyone we know, has a Facebook profile or fan page.

My wife, although relatively new to Facebook, is in love with the social networking aspect of the service. Like everyone on the platform, she has reconnected with her old friends, has connected with her current group and has a direct dynamic connection with our children and grandchildren.

Now, I leave Facebook open on my computer most of the day. For me, it's just one of many communications tools. However, my wife is a Facebook demon. She's on it like glue. Actually, I think she's addicted. For her, it has replaced chat, email and other photo-sharing sites. And that's saying a lot!

Day 1

My wife came into my home office and announced that one of her friends sent her a real-world email saying that they thought her Facebook profile had been hacked. I didn't think much of it. We checked out her profile, messages, etc. and determined that everything looked OK. It wasn't.

Day 3

Several of her friends had contacted her (on and off Facebook) to tell her that they had received messages that they were sure were not from her. Then, I received a Facebook chat message from her while she was standing next to me. The hacker was online!

I challenged the hacker, and he or she told me that the hacker was in my account too. Unlikely, but it sounded fierce.

By the time we got to my wife's computer, Facebook had disabled her account, and she had received a message that her account was compromised and that they were aware of the issue.

That sounded good at the time. But it turned out to be the last good news we heard for a while.

Day 10

Unwilling to use my connections to deal with this and thinking that the experience would make for a good article, I asked my wife to see if she could deal with it. She agreed to try.

Within a few minutes, she told me that she had no idea how to deal with it. There seemed to be no way to contact anyone.

I sent her (via iChat) the URL of the appropriate Facebook security page and told her to follow the instructions. The page clearly says if you think your account has been hacked, "click here." Sadly, it asked for the URL of her Facebook profile, which she did not know. And there was nothing I could do to help her, since her account was gone.

Day 12

After several emails into the black hole of Facebook customer service, my wife didn't know what to do. No profile, no one to call. She was really frustrated.

Day 14

This is actually now a social media nightmare. Debbie Sommers Palmer RIP. Her virtual world no longer exists. Every comment she made, every picture she uploaded, every connection she had, every one of the old friends that she reconnected with—all gone.

Day 15

I decided to go wide with my call for help. I have reached my maximum friend limit on Facebook, and most of my "friends" are in the related fields of technology, media, marketing and entertainment. A large number of them are in pure play Internet. Three Facebook updates yielded several messages of solidarity, but no possible help. Several of my Facebook friends network actually work at Facebook; no answer to the question.

Day 17

From a business perspective, this is the problem with an organization that has over a billion nonpaying customers and no actual meaningful revenue model. How many customer service people would a reasonable person expect them to have? More to the point, the service is free—so it's worth what you pay for it!

But that is not what consumers of Internet services have come to expect. We expect Facebook to live up to our extremely unrealistic expectations of customer retention and customer service. We shouldn't, but we do.

During this particular week, several of my friends' accounts were hacked. Security on Facebook is a real issue. That said, it's a technology problem, and computer people are pretty good at solving those. The bigger problem for Zuckerberg and company is the fact that my wife won't trust a cloud-based social network anytime soon. Will she be willing to invest the time and energy to establish another virtual presence? The hacker victimized her. Then Facebook penalized and victimized her and then abandoned her for days. But just when we thought all was lost ...

Cloud Life Chronicles Part 2: Audrina, the Facebook Ninja

Debbie received a "real world" email that looked like it was from Facebook. It said:

Hi Debbie,

Our systems indicate that your Facebook account has been compromised by cyber-criminals attempting to impersonate you. These criminals often will try to trick your friends into sending them money by claiming that you are stuck in a far away location and need assistance. It is possible that your email account was compromised as well. As such, we have sent this email to all email accounts recently associated with your account. Obtaining access to a victim's email is one of the primary ways these cyber-criminals have been operating. Please change the passwords to any email addresses associated with your account.

There was some additional information in the email with links to webpages that were not helpful without access to Debbie's account. Curiously, the email ended with this request:

In order to regain control of your Facebook account reply to this email with the answer to your security question: What was the name of your first pet?

Thanks,
Audrina
User Operations
Facebook

You want me to email you what? Is it not the first rule of Internet security that you never email anyone personal information about an account? Why would a real security person ask for this in an email? They should have pointed us to a link inside of Facebook where Debbie could enter the information directly into her profile. How could this email be from Facebook Security?

Erring on the side of caution, we replied to the email as follows:

Facebook team,

This is from a person I believe is posing as a member of your security team. She's asking for me to answer a security question that I'm not going to answer. I believe this is the same person who hacked into my Facebook acct. I have not heard from anyone at FB since the first email. I cannot log onto or even get onto FB at this point.

I'd like to know why I'm being targeted twice and what you're going to do about this. Best,
Debbie Palmer

A couple of days later Debbie received the following:

Hi Debbie,

You do not need access to your Facebook account to contact us. We have suspended the account until we can prove ownership. I understand that you are skeptical of our correspondence, but please be assured that you are dealing with a legitimate Facebook representative. If you would like us to reevaluate the status of your account, please reply with the requested information. We appreciate your cooperation.

Thanks,
Audrina
User Operations
Facebook

Audrina's email address is "abuse+njjntt1@facebook.com." We dubbed her the Facebook Ninja.

Over the course of the following week, after a draft copy of "Cloud Life Chronicles Part I" started circulating around the blogosphere, Debbie was contacted by a different Facebook staffer (who asked her to pick a new security question and give the answer—all using email). It was impossible to tell who was a hacker and who worked for Facebook. The email headers were cryptic; the email address had a plus sign in it (not an illegal character, but highly unusual).

Was Audrina stuck in London asking Debbie's friends for money? Is there an Audrina? We'll never know.

Several days later, when Debbie regained control of her account, her profile picture was changed. This freaked her out. How did a different picture become her profile picture? Her Facebook friends didn't believe that she was the real Debbie. It took the better part of the week to get everyone settled down.

The Future

For me, this episode brought several interesting issues into focus.

1. If you have an active user-base that is the size of a large country, does it need a government? I grew up in a town (that had a government) that was in a county (that had a government) that was in a state (that had a government) that was in a country (that had a government). There were four levels of government all overseeing my hometown. Facebook's population is close to 50 percent of the known Internet users. Who represents me?

2. Would I pay for a better version of Facebook? Would you? If so, how much? Is there a frequent flyer program or a paid premium version of Facebook in our future? If I had a better security experience or a set of features that let me automatically back up the things I have on Facebook, would I be willing to subscribe?

3. What does the next Facebook look like? Is the post-IPO Facebook "too big to fail?"

I wanted to ask Audrina, the Facebook Ninja, some of these questions, but alas, like all good, stealthy warriors, she vanished into thin air. She may be gone, but the questions remain. How big is too big? How quickly will we move on to the next new, new thing? Is this just another example of life in the cloud? Only time will tell.

Truthiness, Wikiality, Truth, truth, Facts

Truthiness in a Connected World

What is True? (I don't really want to delve into the metaphysical or philosophical nature of truth. I'm simply trying to help us define and label a baseline.) Are facts true? By definition, they are. George Washington was the first president of the United States. This is a fact. By definition, it is true.

The goal here is to explore the concepts and constructs of Truth (with a capital "T"), truth (with a lowercase "t"), truthiness (as coined by Stephen Colbert), reality, wikiality (also coined by Stephen Colbert) and facts as they apply to our connected world.

The late Senator Daniel Patrick Moynihan once said, "You are entitled to your own opinion, but you are not entitled to your own facts." We can all agree that facts are true, can't we? No, we can't. Here's an experiment my son (and co-author of this section) Jared devised to test the practical existence of fact.

Objective: Explore the way facts, truth, and narrative affect crowds and groups. Observe the relationship between fact, truth, and narrative and apply the findings to the future of social networking, news, and mass media.

Apparatus: Create a video that shows a ball being dropped off a table. Specially color and pattern the ball so that people who are color-blind see it as a distinctly different color than people who aren't color-blind. After having the subjects view the video, they will be asked to describe it.

Hypothesis: Everyone will agree on the fact that the ball dropped. We will call this a "supercluster." The color-blind individuals will say that the ball was one color, and the non-color-blind people will say that the ball was another color. There will also be at least two distinct "truth clusters," those who are color-blind and those who are not color-blind.

My son spends a little too much time in his college physics lab, but I like the physics metaphors we are using here because information seems to travel in particles and waves, just like energy. Here is how 20th-century physics might help you think about the 21st-century information problem we're discussing.

What is true in the 21st century? With a little bit of help from Einstein's relativity, we know that

1. There is no way to distinguish between reference frames and that

2. The laws of physics hold true in all reference frames.

In one of Einstein's famous *gedanken* (thought) experiments, he showed that simultaneity is relative to the observer. Although relativistic effects such as time dilation and length contraction can't be seen from day to day, there are some important implications of Einstein's postulates. No two people observe the same event the same way, and both observers are correct though the numerical values of physical observations may differ. If that's the case, then what actually happens? Who is right and who is wrong? Who can be trusted?

Examples:

*   Barack Obama is the 45th president of the United States of America.
*   North Carolina is closer to the North Pole than South Carolina.
*   The ball (in the aforementioned experiment) hit the floor.

A Truth, according to m-w.com, is "the state of being the case; the body of real things, events, and facts. A judgment, proposition, or idea that is true or accepted as true; the property (as of a statement) of being in accord with fact or reality."

Examples:

*   The ball was blue.
*   The ball was red.

The ball was painted such that color-blind people thought it was a different color than those who saw it who were not color-blind.

What happens when the color-blind observers find out that the ball appeared a different color to non-color-blind observers? A certain number of them will adjust their facts. Will all of them? Should they?

Truth, Narrative Authority, Reality and Wikiality

Stephen Colbert coined the term "truthiness." I like it! This idea will become surprisingly important in the future as the source of news becomes more important. Here's the dilemma:

1. A left-wing cable news show reports some aspects of a physical event.

2. A right-wing cable news show reports other aspects of the same physical event.

3. My friend on Twitter, who claims to have been a witness, reports other aspects.

4. My friend on Facebook reports some mixture, without any qualifying statements.

5. Another friend on Facebook or Twitter writes an opinionated statement about the event.

So, what actually happened? What is fact and what is narrative? Who has more truthiness? What is the reality? What is the wikiality? Are these the same? Which source should I trust? Who has the narrative authority?

The way to solve this problem is to filter and give weight to each source. Marshall McLuhan said, "The medium is the message," but in the 21st century we say, "The median is the message." If you are going to report the news, then you are going to have to be able to make the distinction between fact and fiction, truth and narrative, reality and wikiality. The median, the measure of the central tendency, will become the accepted truth – along the same lines as political philosopher John Stuart Mill's idea of the tyranny of the majority.

Stephen Colbert coined this idea as "wikiality." Urbandictionary.com defines it as "a reality as determined by general consensus rather than cold hard facts. If enough people say it is true, then it is true." J. S. Mill noted that the majority isn't always right; it just so happens to be the most accepted. If we think about each physical event as a single data point, then the lines we draw to connect these lines can be thought of as the narrative. Think about giving meaning to random events as the equivalent of statistical curve fitting.

The accepted narrative will be either the most compelling to us or the one that is closest to the Truth. Maybe the best narrative doesn't even connect all the lines; perhaps it is merely a line of best fit. Regardless of its shape, adding narrative comforts us because it lets us piece together random events—giving life a seeming progression and meaning and the ability to cope with nihilism.

Under these circumstances, it makes sense that the key to narrative authority and Truth dominance is to tap into this narrative addiction and give the consumer/user/audience the most truthful data points or the most compelling narrative.

But what happens when the truth is not the most compelling narrative?

On July 19, 2010, a video was posted by the late Andrew Breitbart to his website, showing video of Shirley Sherrod, the Georgia state director of rural development for the USDA, that showed her making seemingly racist statements at an NAACP event back in March of that year. Foxnews.com was first to report the story that had now splashed through the blogosphere. Later that afternoon, Sherrod was asked to resign via email after a call from USDA deputy undersecretary Cheryl Cook. That evening, NAACP president Benjamin Jealous condemned Sherrod and tweeted that he was "appalled" by Sherrod's actions.

Fact: Sherrod's comments were taken out of context. The full video of the speech, which the NAACP possessed throughout this whole ordeal, reveals Sherrod discussing how she overcame her racism after struggling with it for years since her father's murder.

In this case, the initial wikiality was wrong, as were the facts and narratives given by the news networks as it unfolded. We now know the facts and we can only hope that the facts get out about Sherrod's ordeal. They will, but not to everyone. There will be a fairly large number of people who accept wikiality over reality, after the facts have been revealed. Let's call this residual wikiality. The idea that Sherrod is a racist will continue to be their wikiality because (1) it's the most compelling narrative, (2) ignorance, or (3) lack of effort on their part to continue to follow the story.

Residual wikiality is not new, it has been a fact of life since the dawn of communication. However, in our connected world, those who choose wikiality over reality have two technological tools their predecessors lacked: connected devices that can propagate their wikiality at scale, and an uneraseable, permanent, online record of their wikiality.

Another example takes wikiality to a slightly different place and helps us explore the role of compelling narrative in a less politically charged arena.

This was on the cover of the New York Post on July 26, 2010. Besides A-Rod's frown, it's the beginning of an article on the John Concepcion case. The Post reported

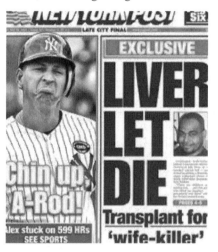

that Concepcion (a convicted murderer who had damaged his own liver by attempting suicide) had somehow made it to the top of the transplant list, ahead of seemingly more deserving people, and been given a liver transplant.

People went crazy about this story. I had a Socratic debate with Judge Jeanine Pirro in the make-up room at Fox 5 about how this happened and what could be done about it. She thought the law should be changed to prevent this from ever happening again, and she wanted to find out how it happened in the first place! Our debate was over how in practice, you could have two systems determining punishment: one legal and one medical. She thought it was easy to do; I was not sanguine with the idea of a committee of medical experts using anything but medical criteria in their decision-making process.

But this story was not a Truth, truth or fact. On July 28, 2010, the Post corrected itself, saying that it could not actually verify that the transplant took place. But how was this correction made? It didn't make the cover. Rather, it was a tiny 2×3 box on the inside flap.

Thus the wikiality of the matter was not and has not been realigned to reality. Can it ever be? No, once a wikiality has been formed, the only way to combat it is with an equal yet opposite wikiality. Sadly, it is almost impossible to create one. As it has been throughout history, the first story, the first headline, the first soundbite is the one that's going to stick. Again, this would not be news except that in a connected world filled with webpages of misinformation and personal truths, nothing can be unsaid and nothing can be unpublished.

So, let's look at trust circles, truth clusters and the way information travels. We'll try to map Truth, truth and facts and think about ways to navigate the body of knowledge as we continue to explore truthiness in our connected world.

About a week after the New York Post story ran about John Concepcion's alleged liver transplant, I ran into Judge Jeanine Pirro in the make-up room again. Remarkably, Her Honor had not seen the Post's retraction and she was unaware that Mr. Concepcion had not received a liver transplant.

The facts didn't change Jeanine's opinion about how the hypothetical situation should have been handled, but it did render our discussion moot. Now, Jeanine Pirro is very smart, and she had no trouble adjusting her worldview to incorporate the new evidence. But there are hundreds of thousands, perhaps millions, of people who did not see the Post's retraction. And they are still operating under the assumption that John Concepcion received a liver ahead of other, possibly more deserving, people.

I'd like to examine how this information traveled around the InterWeb, but first, we need some definitions. Let's define a "trust circle" as our immediate, most trusted sources (friends, relatives, colleagues, thought leaders), our test being that we would trust their opinion over Brian Williams'. (You can use your own benchmark for trust. We trust Brian!) Let's define a "truth cluster" as a group of trust circles with similar beliefs. And just for fun, let's imagine a "super cluster" as a group of truth clusters.

Although this story was initiated in print by a New York Post–credentialed journalist, within a remarkably short time, it was broadcast in the local market using both radio and television. While it was not known to be true, the Post's standards and practices allowed it to print the story as fact. And due to the sensationalist nature of the story, emotions on both sides of the issue instantly came into play. Significantly faster than the story was translated from print to broadcast media, it put the InterWeb in OverTweet. There were dozens of blog posts, tweets, status updates, comments, emails, smoke signals, and carrier pigeons flying all over the place telling all kinds of stories about this story.

There were no facts, just the "truth" with a small "t" printed in the New York Post. However, this truth became a Truth with a capital "T" to some and was assumed to be a fact by others. The idea that a convicted murderer who damaged his own liver while he was trying to commit suicide was somehow put on the top of the liver transplant list made people's blood boil. Trust circles embraced the idea and a huge super cluster of complete misinformation appeared within hours.

In a physical explosion, the energy would have dissipated relatively quickly. And so it did with this story. It's no longer in the news cycle, practically forgotten by all. Except the misinformation can never be unpublished and the rhetoric can never be unsaid. It is all part of the body of knowledge of the InterWeb, the blogosphere, the tweetisphere, Facebookistan, etc. If you want proof, just type any cryptic version of the story into a Google search bar. You'll find hundreds of thousands of mentions in under a second.

We have all seen this kind of behavior before. It's not new. People make mistakes or simply lie all the time. Storytelling is an art form, and there is always plenty of artistic license taken no matter who is telling the story. However, this is the first time in history that we have seen trust circles empowered by instantly scalable technologies.

It is axiomatic that trust circles, our smallest, most inner circles of trusted sources, have always existed. Information Age technology empowers us to overlap all of our trust circles (the one for consumer electronics, the one for religious studies, the one for political guidance, the one for stamp collecting, etc.) and see the results collected in one place. Facebook is a good example. Every one of your Facebook friends is not in every one of your trust circles. They may not be in any, or they may be in a few. Sometimes they are clearly defined by Facebook groups or guests attending a particular event. Other times, they are simply people whom you group together when you are thinking about a specific topic. You assemble your trust circle(s) and you are assembled into the trust circles of others. Facebook and Google+ give you tools to segment your trust circles, but in practice, I don't know very many people who do.

By definition, trust circles are private. You can't "bark" or interrupt your way into a trust circle; you have to be invited in or earn your way in. How you do this is the essence of modern social media marketing. (You can retain my company if you need help with this. It's one of the things we do best!)

Trust circles are the last bit of conscious control any of us have over the contextualized, editorialized transference of information (raw data) into knowledge. As a story propagates through our connected world, truth clusters self-assemble. This, too, is exactly what happens in the offline world.

As we can see every day, super clusters self-assemble around the most compelling narratives. We can track truthiness along the line of best fit (to use a mathematical metaphor) and we can also intuit that super clusters have offline counterparts because we experience them everywhere.

While the structure of human communication (beginning, middle and ending; rising action, climax, falling action) hasn't changed very much over the past 3,500 years, there are two remarkable, nontrivial attributes of narrative today. (1) The most compelling narrative (true or false) cannot be erased from the searchable body of knowledge, and (2) the technology propagates the most compelling narrative (true or false) at speeds that are almost beyond the threshold of our ability to react.

That being said, there is an argument to be made that although the technology makes us faster communicators, it does not make us better communicators. We know this because bad ideas travel just as quickly as good ideas. Facts and truthiness travel equally fast. So it is logical to assume that since good and bad, true and false, right and wrong all travel at the same speed, we should take it out of our thought experiment. But we can't. Speed is, as we will soon see, a huge factor.

Lastly, we have the indelible attributes of the World Wide Web and wikiality to deal with.

If a crowd-sourced wikiality is the most compelling narrative but is not verifiable or fact based, does it become the truth? And if it does, what should we do about it? The answer is, nothing. There is a single word that describes a crowd-sourced, compelling narrative that is not verifiable or fact based: "faith." We do not have any tools, electronic or otherwise, that will reliably, effectively inspire or incentivize people to question their faith. There's no reason to try.

Which leaves us with the only possible tools we can use in the Information Age to help us propagate facts and Truth: filters. Thinking this through and listening to hundreds of learned colleagues tell me stories about truthiness and wikiality, it occurs to me that clear, concise, branded filters are probably the best defense.

Sadly, actual facts and Truth are the victims of the Information Age because filters craft narrative and tend to blanket us in the comfort of the information we want to hear. Fox News vs. MSNBC. Which is truthier to you? Man-made Global Warming vs. Natural Climate Cycles. Which is the more compelling wikiality? Life begins at conception or a woman's right to choose. What does your faith tell you? I could go on forever. But each super cluster believes what it wants to believe, so we don't need to worry about changing minds; we simply need to brand our filters along the lines of best fit.

To aggregate the largest audience and keep it, apply the branded filter that matches the most compelling narrative. Wait … that's what TV programmers do. Yep, that's what they've always done. Is it the "boob tube" or is it "programming to the lowest common denominator?" As we all know, popularity has never been a measure of quality and quality has certainly never been a measure of popularity.

My sons, Brent and Jared (both Millennials), took exception to my idea that social media would empower ordinary people to do extraordinary things. It was their collected contention that I was so enamored with the technology that I failed to see that all of the ambient social media noise was self-canceling. The math, they argued, was simple. They were wrong, of course. The Arab Spring was organized at scale, using social media tools, as was the Occupy Wall Street movement. But mathematical modeling is often inadequate to predict the behaviors of hugely complex systems.

Citing those social media–enhanced revolutions makes me think back to the conditions surrounding the birth of our nation. It occurs to me that the idea that America needed to be independent from Great Brittan was so powerful that it persuaded people who would describe themselves as British citizens to pick up weapons and commit treason.

America the brand was a powerful idea. But treason was a serious crime. Then as now, anonymity was an appropriate tool. Under the nom de plume "Written by an Englishman," Thomas Paine anonymously published "Common Sense" in January 1776. It was an instant best seller. When adjusted for population, it may be the best-selling document in American history. Historian Gordon S. Wood described "Common Sense" as "the most incendiary and popular pamphlet of the entire revolutionary era." "Common Sense" was just a pamphlet, written in the style of a sermon. Would Information Age technology have done a better job changing the minds of British Loyalists? Doubtful.

History, as we well know, is written by the winners. In our time, the Truth (note the capital "T") will be determined by the truth (note the small "t") that makes the greatest number of indelible copies of itself. This is not new, and apparently the transformation of reality to wikiality is not new either. What is new is simply the speed at which all of this is happening. We need tools to interact with information traveling at this speed. And we need better filters to help us sort out and curate the data that are most important to us—without coloring the information so that it is too comfortable.

Is there a way to save facts and Truth in the Information Age? It's easy to predict the evolution of tools that will empower users to find root threads of ideas that are propagating online in near real time. As hand messenger services gave way to faxes, and faxes gave way to email, and email gave way to txt messaging, will we find a way to adapt to the new speed of information? This is a huge opportunity and challenge for all of us who want to be successful in a connected world.

In practice, there has always been a certain amount of opacity between buyers and sellers of information. The lines between salesmen and storytellers and hucksters and liars and cheaters and thieves and hardcore criminals have always been a bit blurry. To succeed in a connected world, you will have to use your best judgment until new technologies evolve to help you fact check. Conversely, those of you who have chosen to align with the dark side of the force can have a field day manipulating information to your heart's content.

The Truth about the Truth of News

Back in March 2007, I received an email from a friend telling me about the plight of Eric Volz. He was sitting in a Nicaraguan jail accused of murdering his girlfriend.

Although this bit of news happened to be true, I was initially unconvinced, and it took me a fair amount of time to figure it out. I chronicled this episode on my blog in an article entitled "Searching For News." The thesis was simple: We live in the Information Age, and unbranded information, even if received from a trusted source, may not be true.

Now we are entering a world (some would argue that we have been in one for a decade) where emails from friends with seemingly real news stories and seemingly real references may be casually passed along and consumed as facts. We've always trusted our friends as good sources of info. We've grown up (even the digital natives and Millennials) in a world of trusted news brands. Why shouldn't we be conditioned to believe what we read if formatted like news is, in fact, news?

This is justthe beginning. User-generated content (UGC) is a staple of our media consumption diet. We love our Facebook walls, Twitter feeds and all the pictures and videos our friends and loved ones make for us. We even like the ones they find and share with us. Sadly, the Moulin of user-generated news (UGN) is seeping under the branded news glaciers we believe will never melt or fall into the sea.

My favorite example of UGN dates way back to the nomination of a relatively

unknown politician as John McCain's running mate in 2007. It spurred all sorts of unsubstantiated rumors, from the baby factor to drug use, drunk driving, guns, political kickbacks, extramarital affairs—the list went on and on. It would make a good soap opera but for the fact that the story is too trite and the characters too stereotypical.

Author's Note: The goal of this chapter is not to push a personal political agenda. It is simply to show examples of the ways information can be manipulated in a connected world and to help you understand the tools and techniques of the trade.

First, there were blogger accusations that Mrs. Palin had covered up her daughter's pregnancy by claiming it to be her own. Some claimed that 4-month-old Trig, who was born with Down syndrome, was Bristol's baby, not Sarah's. As we now know, Bristol was five months pregnant when this story broke, which debunked the rumor. However, the rumor didn't evaporate, and there were thousands of people, if not more, who still believed Sarah was covering for her daughter.

It wasn't just bloggers getting in on the action. So-called real reporters couldn't get enough either. So much so that the fine folks at Bloomberg hosted several rather comical, absurd articles, including the Sarah Palin drunk driving report, which, it turns out, wasn't true. It was her husband, Todd, who was arrested for a DWI, 22 years prior.

In a similar instance, a 2002 story from the Chicago Tribune on United Airlines filing for bankruptcy was mistaken as "new" by Income Securities Advisors Inc. and subsequently picked up by Google News and, you guessed it, Bloomberg News. The story effectively tanked United Airlines stock, causing a 76 percent drop in share value. All this just two years after United emerged from bankruptcy protection.

One of the most popular of the Sarah Palin stories was the infamous American flag Bikini/AK-47 photo. This Photoshopped masterpiece spread through the Web like wildfire, so much so that CNN reporter Lola Ogunnaike commented that Palin "looks good in a bikini clutching an AK-47, but is she equipped to run the country?" Too bad it wasn't the governor and the gun was not even an AK-47!

Another piece of Palin intrigue was the open letter written by Wasilla, AK, resident Anne Kilkenny, who had supposedly known Palin for years. The letter highlighted Sarah's rise to power and her actions along the way. Kilkenny was objective, and honest in her assertion that she herself has sparred with Palin in the past, specifically over the banning of books in the town library. This letter was picked up for publication by The Nation, The [Illinois] Daily Journal and the Anchorage Daily News. But, as with the Eric Volz story, I received an email about this letter from a friend in Upstate New York, who told me that he had received it from two different people that morning. Were his sources credible? Was this letter real?

My imperfect solution was to check the letter out on urbanlegends.com[3] and snopes.com.[4] They said the AK-47/bikini pic was fake, the list of banned books was fake and the letter from Anne Kilkenny was partially true. The open letter from Ms. Kilkenny, a lifelong Democrat who supposedly attended every city council meeting during Palin's first year in office, was even featured as part of an article in the New York Times.[5] Partially true?

Media professionals have slid down the slippery slope of journalistic integrity much farther than I could ever have imagined. Some of our most trusted news sources are using UGN as source material and have no more ability to check their facts than average Internet users do.

Mistakes happen. UAL was serendipitously taken out to the woodshed and their share price with it. It could have happened to anyone. But having a CNN reporter think, for a second, that the picture of a woman in a bikini holding a rifle could actually be Governor Palin and reporting it as such is truly inexcusable for a professional news organization. Anyone with a minute of training could tell you that the weapon was not an AK-47 and that the picture was most likely "fun with Photoshop."

In these times of ubiquitous communications tools, video production, audio production and graphic arts capabilities, it is incumbent for professional news media outlets to exercise above-average judgment and demonstrate a higher standard of editorial decision making. But I would not hold my breath until that happens.

Big Data and Life in the Cloud

You Do Not Need a Website!

I recently got a call from a prospective client. They wanted me to assist them in building a website for their new online business. It was quite a conference call. There were all kinds of really smart people involved. The demo screen shots they created were stunning. Pretty pictures, awesome usability, excellent site architecture. I was impressed. As the presentation continued, they showed me all of the features the site was going to have: social web stuff, news feeds, interfaces to Facebook, Twitter and LinkedIn, data pulled in from all kinds of interesting places—you name it, it was on the list.

Only one problem: They didn't need to build a website. They needed to build a database!

It is remarkable how many people still come to the online world believing that a website is the central core of their operation. Nowadays, that seems as archaic as saying that a television show is the central core of your operation. We're in the Information Age. Everyone is in the customer relationship management (CRM) business, and that starts with a database.

Now, you might want to build a website that allows your customers to access your data (content, video, audio, text, graphics, pictures, etc.). But you will also need an easy way to supply their needs on smartphones, app phones, cell phones, landline phones, PCs, notebooks, slates, tablets, iPads, iPods and a whole host of other devices that are out there and those that are yet to be invented. You're going to have an App vs. WAP debate, and sadly, you're going to decide that you need to create both consumer experiences. You might also have to supply your video to broadcast television, cable television, satellite television, on-demand systems (online and offline), IPTV and even lowly YouTube. Audio may go to traditional radio, podcasts, online distributors, iTunes, Rhapsody, Pandora and a thousand others. This list of deliverables is not endless, but it is frighteningly long and getting longer every day. The data won't come out of a website; they will come out of a database that can be accessed by a plurality of devices in a plurality of ways.

People who are in the "website first" business get their knickers in a knot when I bring up this nontrivial strategy shift because they use all kinds of databases when designing a site. There's the content management system, the image server and the video server, to name a few. All of these are databases.

But the difference here is flexibility. If you create a website first, you are committing your thinking to optimizing and specializing content for one type of online delivery. It simply doesn't make sense to do that when you know full well that you have to make your content available and optimized for whatever experience consumers popularize. I don't want to even try to guess which new, bright, shiny object will win the hearts and minds of consumers on any given day. New devices are created all the time. Manufacturing cycles are getting shorter, and you really can't stay ahead of them.

If you're starting a consumer-facing services business, you should seriously consider starting with a robust content management system, good CRM, and a best practice integrated transaction engine and accounting system. Designing the digital infrastructure and database first will give you the best opportunity to maintain margin and manage your digital business as consumer consumption patterns evolve. And it won't compromise your website design in any way. OK, I lied. You do need a website, but it's really just a way to view your data.

Targeted, Addressable Advertising

There are several marketers and third-party marketing companies that swear by the concept that the more relevant a message is, the more effective it is. This assumption, which makes intuitive sense, is the basis for a huge industry in targeted, addressable advertising.

For our purposes, targeted advertising is defined as messages that are aimed at a specific demographic or cohort, such as "Teens 13–19" or "Adults 18–49" or "Two-headed Martians that eat their young." You can select any type of audience to target.

Addressability is the ability to specify where a message should be delivered. Just like "123 Any Street, Any Town, 12345-1234, USA" is a physical address, every connected device has an electronic address which can be collected into a database and used to distribute messaging.

Targeted and addressable advertising advocates believe that a significant amount of traditional advertising is wasted. Their guiding principle is based upon the most famous quote about advertising:

"Half the money I spend on advertising is wasted; the trouble is I don't know which half." —*John Wanamaker, US department store merchant (1838–1922)*

If Mr. Wanamaker's famous quote is true, certainly data science should be able to provide an answer.

The problem is that this quote and the idea that 50 percent of a brand/lifestyle advertising budget is wasted (as opposed to a direct response or a call to action advertising budget) is intrinsically false.

The targeted and addressable crowd believes that if 10 million people simultaneously watch one message, and it is only relevant to 5 million of the viewers, they can correct for that "wasted" 5 million viewers by sending relevant individual messages to 10 million people simultaneously instead.

Even if the technology to accomplish this were fully functional (which it is not), this kind of messaging does not work for advertisers who need scale to sell their goods and services. Select, highly relevant ads may achieve slightly higher efficacy metrics, but come at the expense of buzz. And getting enough buzz to have *SNL* or Jon Stewart call you out or to be the topic of "water cooler" conversation simply cannot be achieved without scale.

The additional problem with targeted and addressable advertising technology is that it is priced so high that the cost per acquired result (click, action, acquisition, etc.) usually does not justify the budget allocation. This is a generalization, of course. There are many, many cases with direct response or call-to-action advertising where precise targeting yields extraordinary results. You just can't think of targeting and addressability as "silver bullet" solutions to advertising messaging in the 21st century. They are not a "cure all" solution.

Inductive vs. Deductive Reasoning: The Chicken and the Egg

Imagine that you are a professional observer with a magic telescope. Like ordinary telescopes, it lets you see things that are far away, but this one is attached to a magic screen, so I can use it to tell my story.

On your magic screen, you see a farm in a distant land. Each morning you observe the farmer feeding his chickens a measure of food, and each evening he returns to harvest about a dozen eggs from the coop.

One day you notice that the farmer has doubled the amount of food that he is feeding his chickens, and lo and behold, the output of eggs virtually doubles. Since you have chosen to scientifically observe this farmer's behavior, you and your team of experts conclude that (1) the farmer has figured out or (2) the farmer has somehow learned that doubling the amount of food will yield approximately double the number of eggs. You go on and theorize from what you have observed that the marginal gain from egg production exceeds the marginal cost of chicken feed, and the farmer is now running a more efficient and more profitable operation. You decide to call the place "The Egg Farm."

What you failed to realize (because the magic screen does not let us hear conversations or communicate with what we are observing) is that doubling the amount of food a chicken eats also results in fat chickens. And since you didn't know that eggs were in abundance in this farmer's world, you could not know that market pressure had pushed the marginal cost of producing eggs past the point of profitability. However, this particular farmer's business advisors had told him that the price for chicken meat was on the rise, so he should fatten up his chickens, sell them, and become an emerging media pundit.

He tells his advisors that he doesn't need to become an emerging media pundit because he has not taken any of the eggs to market since he started doubling egg production. He has put them in an incubator in another building and transitioned his business into a "Chicken Farm."

People who are closer to the situation could have easily deduced what was happening from their more detailed observations. But on our magic screen, we only saw what we saw, so we were forced to reach our conclusions using inductive reasoning. This is never a good thing. After all, through our magic screen we still think this guy is in the egg business!

The key take-away from this story is simple: Information is not knowledge.

In the age of Big Data, data scientists can make numbers say anything. To translate information into knowledge, it needs to be contextualized. In a connected world, data are the fundamental building blocks for Reductionism. For example, does it really matter why a certain website is the most popular of its type? It might if you were programming it, but it may not matter if you're purchasing advertising on it. Reductionism is rampant in the age of Big Data, and to be truly successful, digital leaders need to become Big Data fluent.

The Role of IT

It seems as though in every meeting I take, someone in marketing or management delights in vilifying the company's IT (Information Technology) department. While it is true that IT departments are not always filled with "yes" people, they are generally not the enemy.

Back before computers, management used to keep the books of the company in a safe and lock it every night. To this day, bookkeepers are not allowed to use pencils to keep ledgers, journals and registers. It is all done in pen (black and red). When a mistake is made, a special entry called a "reversing entry" is made to correct it. As Professor Horowitz once admonished in my Accounting 101 class, "Doctors can bury their mistakes, accountants cannot!"

Fast forward a few hundred years, and management asked IT to secure the books of the company as the business processes were computerized. As you know, financial software is some of the best-protected, most secure software. The systems are, by definition, closed.

Fast forward to the age of email, blogs and social media where the public Internet and World Wide Web are, by definition, open.

Closed systems and open systems can function in the same world, but it is really hard to make them work together.

So, while it sounds like the people in IT only know the word "no," it is important to remember that they are being asked to do a very difficult task: keep the closed systems secure while granting limited access to the open systems you need to work. Considering the mission and the stakes, I'm surprised that anyone who works in traditional IT even attends meetings with marketers.

Google Drive: Business with a Single Point of Failure

You could think of Google Drive as Google's answer to Dropbox or Microsoft's SkyDrive or any one of a dozen cloud storage systems, but it's not. Google Drive is so integrated with Google Apps for Business, Google Docs and Google that it is destined to become the seminal point of "the" paradigm shift to business in the cloud. Or, as I like to think of it, business with a single point of failure.

The allure of Google Drive is so great that at my company, we thought seriously about dumping Microsoft Exchange and Microsoft Office (specifically Outlook) and taking our relatively insignificant Gmail accounts and making them central to our business.

Google Apps for Business has some very attractive features. It's relatively inexpensive, it's integrated perfectly with Salesforce and now it's integrated perfectly with Google Drive. Seamless integration of all of our documents, shared Google Docs and spreadsheets, client folders, scanned documents, email, calendar, etc., is like the ultimate geek fantasy.

> Angel Geek: "Whoa. Hold onto your hard drive. This makes no sense at all."
>
> Devil Geek: "Why? What could possibly be bad about having all my stuff in one place?"
>
> Angel Geek: "Duh! How about a single point of failure for my entire business?"
>
> Devil Geek: "Yes, that would be bad."

Let's look at it another way. There's a reason that a balanced portfolio of investments requires you to spread your money around. It allows you to survive an unexpected, catastrophic disaster with any one investment. A well-diversified portfolio of investments, properly hedged, is the gold standard for professional hedge fund managers. In fact, the term "hedge fund" describes the business model—a hedge against downside risks.

I consider information "the" currency of the Information Age. And right now, my business information is in my version of an Information Hedge Fund. I have documents stored locally and in several different clouds. My financial data are backed up on physical media, in a different cloud and locally. You get my point. I've got information, the currency of my business, stored in all kinds of places, some safer than others.

So what would possess me to put all of my InfoCurrency in one place? What is the upside? There is none. It is a remarkably stupid idea. Single-threading your business through a single point of failure is like putting all of your money in one pretty good investment and hoping for the best. You could make a profit, but if anything goes wrong, you lose. Of course the things that can go wrong are completely out of your control, just like financial investments. You have no say over new regulations, lawsuits, and market conditions. You're just hoping that your investment advisor picked the right investments.

Will professional IT guys use Google Drive and Google Apps the way amateur IT guys use them? Of course not. The pros know how to hedge against single points of failure. But 99.99 percent of Google customers are not IT professionals; they're just people. And people will not understand the inherent risks of a single information investment because they don't think of information as currency.

The concept of a single point of failure is not specific to Google Drive. If you're a Microsoft-oriented business (Office 365 and SkyDrive), your single point of failure will be at Microsoft. Same/same.

That said, you now know that information is "the" currency of the Information Age, and you are not going to be lulled into putting all of your data in one place. You are going to figure out how to spread your data around into multiple safe-to-mostly-safe places, so if a crash or a cyber-security breach or something bad that none of us is smart enough to foresee happens, your InfoCurrency is safe and your business survives.

Privacy

An Unreasonable Expectation of Privacy

People who are schooled in the art of the Internet or those who understand how computers work know that any time you interact with a computer, your actions can (and probably will) be logged. The level of detail depends upon who is doing the logging, but it is not unreasonable to expect your log-on time, log-off time and keystrokes to be recorded. This, of course, includes everything you type as well as the address of every Web site you have visited and what you did there, every word of every unencrypted email you send and receive, every search you do, everything.

Who would or, more importantly, who should care about living in such a world?

Two-way, interactive communication over vast networks theoretically offers limitless potential for good. The more information you allow "the system" to have about you, the more the system can adapt to your needs, wants and desires. The disciplines of behavioral targeting and collaborative filtering, when combined with data mining and data warehousing technologies, can in theory offer a world of information and entertainment in ways that test the limits of one's imagination.

That's the sales pitch. The reality is very different. Once "the system" knows you, someone must ask the question, "Who knows the system?"

This question didn't even exist in the past. (And it doesn't really exist for people who spend most of their time offline.) In the offline world, statisticians make a very nice living aggregating data and predicting what populations might do. They don't mind using anonymous data because it is accepted dogma that statistical analysis does not attempt to predict what specific individuals within a population might do. However, in the online world, there are no populations. There are people. Populations may do things, and there may be some value to trend analyses, but those aggregations are simply constructs for management. The actual data exist, and they exist on a "big brother" basis.

What makes this thought experiment interesting is the idea that while the predictive power of aggregated data can be "stolen," it has very little malevolent usefulness. On the other hand, aggregated behavioral data tied to an individual and gathered without the express permission (or knowledge) of that individual have so many malevolent uses they cannot all be listed.

We are quickly evolving into a technocracy where the expectation of privacy is unreasonable. If you sit down at a computer, you are "online." Is what you type protected? What if you need the help of your community of interest?

What if you must ascribe to a collaborative filter set to enable your discovery of new content? What if you want to take advantage of the recommendation engines from retailers? What if you want to become part of a tag cluster or social network? Can you reasonably expect privacy?

Most people would say their privacy has been violated if someone obtained and distributed information about them in a way that was inconsistent with our social norms or the reasonable expectation of privacy. Obviously there are legal definitions of personal privacy, although the current laws do not define current technology as well as they might.

In the offline world, people "do" life in private. Some people even do business in private. That's the role of printed money in our society. The local street vendor does not give you a receipt for the pretzel you bought on your way back to the office. Will he report the transaction to the government? Will he pay taxes on the income? Will he log the purchase and make the data available to your nutritionist? Your doctor? Your significant other who has put you on a low-carb diet? Of course not. One can think of literally hundreds of transactions that ordinary people complete that they would "prefer" to do anonymously.

"Anonymously"—an interesting word. But not one you should grow attached to. As we harness the power of technology to make the doing of life better, we will be moving toward a world where the expectation of privacy is totally unreasonable.

Apple and Google Spy Planes Are Looking at You!

Apple and Google are using military-grade spy planes to map your back yard. That's the sensationalist headline. I like it. It's really scary and it's perfect for FUD-mongering. (FUD is an abbreviation for fear, uncertainty and doubt.) Similar headlines have everyone from privacy advocates to private citizens up in arms. After all, the 17th-century proverb "Good fences make good neighbors" has no meaning in the age of spy planes. According to some reports, the resolution of the spy cameras is so good that you can see objects as small as four inches. Other reports foretell of the capability to see through skylights and open windows. All kidding aside, it's pretty creepy.

But whether this particular use case creeps you out or not, the creep factor is not the issue. The issue is framed by the question, "Is there any reasonable expectation of privacy in the 21st century?"

I submit that this is the perfect time to start a dialog about what privacy is, what it is not and what rights we should reasonably expect.

To start this dialog, we need to understand the full nature of the technological capability we are up against. Apple and Google using military-grade tools to create better way-finding tools for maps and other applications is actually pretty benign. Want to think about surveillance? Great. Let's think about it the right way.

Within a couple of years, file storage will be inexpensive enough and computers will be small and powerful enough to create work environments that are under Star Trek–like audio and video surveillance 24/7. This might mean, for example, that every conversation you have at your job, no matter where it is and no matter who it is with, will be recorded and analyzed. The computer programs will be trained to understand what you are saying, syntax, context and to flag anything that is deemed inappropriate or un-business-like. Forget the totally lame "This call is being recorded for quality assurance." They're going to record everything.

Does this sound Orwellian or paranoid? It may. But unless you assume that the technology is already in place, you can't start thinking about writing rules and laws or even setting guidelines. I promise you, the capability to spy on your every word, movement, email and electronic transaction is already in place. The only thing that is missing is a cost-effective way to deal with the explosive amounts of data that this type of surveillance generates, and that capability is easily within site.

Right now, companies asking you to store your music and movies in the cloud are bombarding you. That's a good use of cloud technology, but you know what cloud technology is really great for? It's great for big audio and video surveillance files. In fact, it's the best place for them, because cloud storage is easily accessed by cloud computers, and cloud computers have the capacity to crunch extremely large files.

So, now that you are thinking about an invasion of your privacy so deep that it is actually disturbing, what should we do? How should we start to sort out the issue? What are the questions you want to ask? What are the guidelines you will insist upon?

If under some circumstances you might be willing to be observed in this way, who should have access to the data? Should anyone be allowed to correlate them to other data, such as medical or financial records? I've often wondered how much money could be saved (and earned) by simply using EZ-pass data to issue speeding tickets. If the speed limit is 55 mph and you go through two toll booths in less time than it should take you at 55 mph, you are unequivocally guilty of speeding. Why don't they just put a few high-speed EZ-pass lanes on the highway, fire the cops and send tickets in the mail? It would be easier, cheaper and a much better profit center.

OK, now, expand this to everything you do? No matter how good your imagination is, no matter how much science fiction you've read, you're going to come up short on the ways this kind of surveillance data can be obtained and used.

So let's take some action. Contact your elected officials and tell them how you feel about privacy. Your privacy. My privacy. Everyone's privacy. What are we as a society willing to give up for the quality of our digital enjoyment? It's "the" question of our time.

Do Not Track ... Really?

A couple of years ago, the FTC proposed a "Do Not Track" button. The idea is really simple: Add a simple way for consumers to opt out of Web tracking systems. In practice, it's a little less simple: Open your browser preferences, select the privacy tab and look for a checkbox that says something like: "Tell websites I do not want to be tracked."

Having done so, a reasonable person might assume that he or she was now free to browse the World Wide Web without leaving a trail of breadcrumbs for Web profilers, unsavory advertisers and custom content providers. Not exactly.

As is often said, "The road to hell is paved with good intentions." According to the *Wall Street Journal*, "The new do-not-track button isn't going to stop all Web tracking. The companies have agreed to stop using the data about people's Web browsing habits to customize ads, and have agreed not to use the data for employment, credit, health-care or insurance purposes. But the data can still be used for some purposes such as 'market research' and 'product development' and can still be obtained by law enforcement officers." Seriously?

First, let's look at the administration's framework for the Consumer Privacy Bill of Rights:

- Individual Control: Consumers have a right to exercise control over what personal data organizations collect from them and how they use them.

- Transparency: Consumers have a right to easily understandable information about privacy and security practices.

- Respect for Context: Consumers have a right to expect that organizations will collect, use, and disclose personal data in ways that are consistent with the context in which consumers provide the data.

- Security: Consumers have a right to secure and responsible handling of personal data.

- Access and Accuracy: Consumers have a right to access and correct personal data in usable formats, in a manner that is appropriate to the sensitivity of the data and the risk of adverse consequences to consumers if the data are inaccurate.

- Focused Collection: Consumers have a right to reasonable limits on the personal data that companies collect and retain.

- Accountability: Consumers have a right to have personal data handled by companies with appropriate measures in place to assure they adhere to the Consumer Privacy Bill of Rights.

At first glance, some of these ideas, such as individual control, transparency, and security, are good. Others, such as respect for context and focused collection, sound good but will be very, very hard to put into practice.

Does the United States need a Consumer Privacy Bill of Rights? Absolutely! We need to figure out how to govern our budding digital democracy. Is this the right way to approach it? Probably not.

FTC Commissioner J. Thomas Rosch has posted a .pdf you should read entitled THE DISSENT: WHY ONE FTC COMMISSIONER THINKS DO NOT TRACK IS OFF-TRACK Rosch: Concepts to Guard Online Privacy Have not Been Properly Vetted. Here are some of the relevant paragraphs:

First, there are a number of consequences if a consumer adopts a do-not-track mechanism. To begin with, a consumer may sacrifice being served relevant advertising. On a related note, there is academic research suggesting that in order to compensate for the loss of the ability to track consumer behavior and the associated ability to serve relevant advertising, advertisers may need to turn to advertising that is more "obtrusive" in order to attract consumers' attention.

Consumers may also lose the free content they have taken for granted. Not only could consumers potentially lose access to free content on specific websites, I fear that the aggregate effect of widespread adoption by consumers of overly broad do-not-track mechanisms might be the reduction of free content, free applications and innovation across the entire Internet economy. Beyond that, consumers may forgo the reported ability to earn commissions from "selling" the right to track their behavior or allow the use of their personal information.

I also wonder whether an overly broad do-not-track mechanism would deprive consumers of some beneficial tracking, such as tracking performed to prevent fraud, to avoid being served the same advertising, or to conduct analytics that foster innovation. Concerns have been raised that do-not-track mechanisms also may have the unintended consequence of blocking tailored content, in addition to advertising.

Commissioner Rosch is spot on. Sadly, he's not getting much traction with his reasoned approach to the issue. Our elected officials see privacy as the "gift that keeps on giving." You can scare the hell out of people, then launch a campaign focusing on shutting down the evil Internet people. It's truly the perfect issue for those who prey on the fearful.

What can we do? First, contact your elected officials. We live in a republic where our elected officials speak for us and make our laws. They need to understand how you feel about privacy. What do you want? Tell them.

What if you don't know what you want? Admitting that takes courage, and by the way, it is the right answer. This is new territory. We have never been here before. It is impossible to imagine all of the things people will do with data in the age of Big Data. But knee-jerk political reactions and fear mongering are not a reasonable course of action. Like so many things in our modern world, this very important issue cannot be solved with sound-bite politics. Privacy, data collection and what data can and should be used for is "the" issue of the Information Age. It would be great if we could all give it the respect it deserves and start a Socratic, earnest, serious dialog. Information is the currency of the Information Age, and we should treat it that way.

Google = Skynet ... Yikes!

The Shelly Palmer School of Connected Living has one primary thesis: "Technology is good." I believe that all technological progress is good and that the story of the evolution of mankind is inextricably linked to the story of the evolution of our technology. We are tool builders, and we are tool users. It is, in large measure, what separates us from virtually every other species in the known universe.

I also acknowledge that "Technology is good" is an optimistic point of view. I am, by nature, an optimist. I believe in lifelong learning, and I aspire daily to the joy of striving to realize things that exist in our imaginations. It may be one of our higher callings; it is certainly one of mine.

So I am usually one of the guys who says things like, "Guns don't kill people, people kill people." Which is my way of acknowledging that firearms are simply tools to help us throw rocks faster and more accurately, and if you need to throw a rock, it's probably better to throw it faster and more accurately.

This argument can be extended to less emotional subjects like the Sony Betamax case or the more recent (though seemingly ancient) Grokster case, both of which ended up with the court deciding, and I'm paraphrasing: "Technology good ... people bad."

"If God intended us to fly, he'd have given us wings." Yep. I totally agree. God (please use your politically correct deity; this article is not about science vs. religion) gave us brains that saw birds and imagined what it would be like to fly. The same deity gave us thumbs, manual dexterity and the ability to create tools that enabled us to have wings. We fly because we are genetically gifted to do so. (You can decide how those genetic gifts were bestowed. Like I said, it is not the point of this writing.)

The point is that technology is woven into the fabric of our lives, and in every case, in every civilization (past and present), it defines how we interact, how we live, how we work. It defines everything about us, including the epochs and ages of our past.

The reason for my huge pro-technology buildup is that I am about to write something that is so out of character, so remarkably against one of my strongest personal axioms, I have to talk myself into writing it. Google has gone too far.

On March 1, 2012, Google consolidated the privacy policies for 60 of its products, creating the singularly most significant database of the Information Age. The aggregation of these data empowers Google to correlate and contextualize our thoughts, aspirations, actions, physical locations and timelines for the basic processes of the doing of life.

I don't think any single thought about the aggregation of data or the use of technology has ever made me as uncomfortable as this announcement. On its best day, with every ounce of technology it could muster, the U.S. government could not know a fraction as much about any of us as Google does now. But now is not what I'm worried about. I'm not even worried about this decade. At the current rate of technological change, taking into consideration the amount of information we are creating about ourselves and adding in the computational power that will be available in about a decade, Google will equal Skynet circa 2022.

This is a guess. Of course, it could be sooner, but it won't be later. What do I mean by Skynet? First of all, get your Terminator lore together, but then just imagine a database that could automatically determine what you are most likely going to have for dinner after your bowling league Tuesday night, where you are going to have it, whom it will be with, whether you are feeling good or have a cold, if you and your wife are fighting, how your day was at work, what you are thinking about buying, who is helping you with your decisions about it, what chronic illnesses you are dealing with, what meds you are on, etc., etc., etc. And this isn't even the scary stuff.

What scare me are the advance of analytical tools and the existence of yet-uninvented ways to manipulate data for good and, inadvertently, for bad. I'm not worried about bad people doing bad things. That is the nature of our world, and generally it is easy to identify bad people who do bad things. I'm worried about the good intentions that pave the road to hell. I can't speculate about how our near-term-future, data-dependent culture will be negatively affected by the law of unintended consequences. That's because so many of the vocations and avocations that will be impacted have also yet to be invented. I just know that there are at least as many ways for things to go wrong as there are for things to go right.

The sky is not falling, and this is not a sensationalistic, FUD-mongering exercise (Fear, Uncertainty and Doubt). It is an admonition that the time has come for learned colleagues to start a Socratic discourse about what parts of the genie need to stay in the bottle and what parts can be let out. Imperfect metaphor? I don't think so.

This is a very complex problem, and we are going to need very simple ways to describe it. Skynet can't win—at least not in the world I want to live in. Let's get ahead of this while it's still just the subject of the occasional rhetorical blog post, because no matter what anyone tells you, the world of Big Data is never going away.

Mobile

There are almost as many mobile devices as there are people in the world. In practice, most people have more than one device, so the statistic is silly. What isn't silly is how significantly mobile connected devices have changed and are continuing to change the world. I don't know if you can use mobile social media apps to sell toothpaste, but I am completely sure you can use them to overthrow governments.

Even with its remarkable penetration, mobile has been hard to monetize. Google searches, which are remarkably profitable when done by PC users, are far less valuable when done via mobile device. Google (which owns the Android operating system) is the most ubiquitous mobile operating system worldwide, and it has not cracked the mobile code.

At this writing, Facebook is doing a terrible job with mobile advertising and mobile commerce. Search and social may not be the best way to make money with mobile.

However, mobile devices offer exceptional opportunities for companies and brands that use them well. I won't call out any particular app here, but there are literally hundreds of companies that are doing a great job implementing the following strategy: *Create an omni-platform experience that allows users to seamlessly interact and transact with our brand.*

The most important attribute of this strategy is the concept of omni-platform distribution. A website is not an online strategy, and an app is not a mobile strategy. Every business that wants to succeed in a connected world needs to create and maintain a best practice consumer relationship management (CRM) database, content management system (CMS) and Digital Asset Management (DAM) database and craft an omni-platform system that allows users (consumers and workers) to access the appropriate data on any device.

One of the best ways to think about your mobile strategy is to ask yourself the question, "Can a consumer instantly, painlessly, seamlessly interact or transact with me whenever they want to?" You've probably already experienced this. Perhaps you've been discussing something at dinner with friends and someone says, "Oh, I just finished the greatest book called _____." You type anything that even resembles the title of the book into Amazon.com or the Amazon app (Amazon will understand what you type, even if you're all thumbs) from your mobile device while you're having the discussion and put the book on your wish list or one-click purchase it. It could not be easier. This is the gold standard; I challenge you to do better.

Transactions: eWallets, Near Field Communication (NFC)

Mobile wallets and mobile transaction tools are some of the most exciting emerging technologies in our connected world. The idea of using a mobile device as a credit or debit instrument makes bankers and merchants smile. There are a fair number of significant hurdles that must be overcome before this technology becomes widespread. First and foremost is security. But convenience and standardization are also sizable obstacles on the road to eWallets and mCommerce.

The benefits to businesses are obvious: better CRM, easier accounting, greater institutional knowledge. The list is long and wonderful. The benefits to consumers are less obvious but significant. Most transactions are done in cash. Credit card debt is already bandied about as the next burstable bubble, and debit cards are stigmatized as down-market financial instruments.

All of this will get worked out over time. How much time? If Kurzweil, Moore and Metcalfe are used to calculate it, the answer is: very, very soon.

I could, and probably should, write an entire book on the soon-to-be-explosion in eWallets, mobile transactions and mobile banking. Suffice it to say that mobile commerce is more than a trend; it is a behavior-changing technology that needs your immediate close attention and leadership.

MoSoLo: Mobile, Social, Local

There is a concept in mobile commerce known as MoSoLo or LoMoSo or SoLoMo. All of these are words made from the first two letters of Mobile, Social and Local. No matter how you say it, the idea is the same: Good things happen when you combine mobile devices with social media and location-based tools.

This is absolutely true for certain types of businesses. But like so many things in our connected world, it is not true across the board. Location-based car-service apps are extraordinary. Location-based, social, game-ified, mobile apps for bar hopping or restaurant finding are awesome! A news app that knows where you are and brings you a hot article about the local school board in a town you're just visiting? Not so awesome.

A good way to think about the Lo in MoSoLo is to structure your tactical execution around the concept of hyper-personal, rather than hyper-local. Data, information and knowledge delivered "locally" are probably less meaningful than the same package delivered "personally." The trick is to use the location to add value, not just for its own sake.

As we are about to see, local—especially in the context of its value to publishers of advertising—is not what it's cracked up to be.

The Future of Retail

The Complete Myth of Local Advertising

Harry emailed me the other day. He's become a loyal viewer of my television show, and he figured he'd email me and ask me if I could send him info about Web marketing for his carpet and floor covering company. I wasn't sure quite what to send him, so we set up a call.

After pleasantries were exchanged, Harry cut to the chase: "Can you get my website to the front page of Google?" Really, this is an exact quote. "… the front page of Google?" What Harry wanted was obvious, but his choice of words betrayed anything other than superficial knowledge of what he wanted from me.

To be polite, I suggested that before he did the SEO and SEM necessary to accomplish his goals on Google, he might want to think about what his business goals were. He told me that his website was created for him for free by a company that now wants to charge him money, but he thinks they are asking too much.

I told him I thought his website was worth exactly what he paid for it and suggested that he take it down and put up a nice splash page with some pictures of carpeting, the locations of his stores with links to Google maps and his phone number. Then he asked me, "Will that get me to the front page of Google?"

At this point I was fascinated with the conversation, so I went into my standard explanation about business goals, like selling more carpet and floor covering. I spoke about conversion metrics and how he might measure the success of his web marketing efforts. Driving foot traffic over the doors of his two retail locations, etc.

"How much will this cost?" asked Harry. I answered, "It won't cost you anything. It will make you money." Harry did not understand. We discussed the investment he would need to make in a comprehensive marketing plan for his business and spoke about workflow, execution and the differences between advertising, marketing, sales and public relations. After a 10-minute lesson in 21st-century retail marketing, Harry asked me, "Will that get me to the front page of Google?"

Finally, I asked him how much he thought he should spend to create a website that would increase his business. "I don't know," he answered. "The one I have was free."

When I reiterated that his free website was probably hurting his retail business rather than helping it, he asked me for some free suggestions that he could implement for free that would—yep, you guessed it—get him to the front page of Google.

This is a real conversation that actually took place. I've changed the owner's name, but other than that, this is exactly how it went. Let's review:

- A retailer with two doors, one in Manhattan and one in Brooklyn.
- A website that was created for him as a promotion by a template-using website company with the hopes that he would eventually pay.
- A business owner with absolutely no clue as to how advertising, marketing, sales and PR work in the 21st century for local retail businesses in his vertical.
- A business owner with zero aptitude and zero headcount to implement even the simplest technological solution.

This week, I have seen about 20 pitches from companies offering hyper-local and location-based solutions targeting local advertisers. Next week I will probably see 10 more. Hasn't anyone spoken to Harry?

There is no incremental local retail advertising to be had. The money simply isn't there. If a local company is big enough advertise, it is already doing it. If it is not big enough to advertise, there's a reason. The myth of local advertising is that it exists at all. It simply does not.

There is no version of the world where Harry's business is worth going after, or worth taking. He will require three times the amount of customer handholding that a customer three times his size would require. He will never spend enough to justify speaking to him. He will torture you for every dollar he is asked to spend because of how hard he has to personally work to make the dollar in the first place. Harry is a real person with a real business, but he is not a growth opportunity for a technology-driven hyper-local advertising business. Harry is not a growth opportunity for anyone—not even for himself.

You can't go door-to-door to find lots of Harrys. You can't afford the customer service. You can't expect him to use a dashboard without training. It will cost you so much money to acquire Harry as a customer, you could never get an ROI on the customer acquisition cost.

Next time someone brings you a new business model and talks about local advertising as the market, look up a local carpet and floor covering retailer with a couple of doors and $1.5 to $2 million in gross sales. Ask the proprietor about how you can help him, and don't be surprised if he asks you if your technology can "get him on the front page of Google."

Now, for every Harry—and there are lots of them—there is a new use of technology that literally threatens "business as usual" in very real ways. What will happen to Harry when some smart kids on the West Coast see an opportunity to disintermediate his business by altering time-tested wholesale and retail distribution channels? Let's have a look.

Subscriptions Will Change Everything

One by one, almost every product that can be offered by subscription is being offered online. Diapers at diapers.com, drugs at drugstore.com, food at freshdirect. com and razor blades at dollarshaveclub.com. This is more than a fad. It is a trend, and unlike some trends, the effect subscriptions will have on brick-and-mortar retail is easy to calculate and hard to ignore.

At this writing, I have it on the highest authority that approximately 10 percent of the diaper business in the United States is done via online subscription services such as diapers.com and amazon.com. Ten percent does not sound like a big number, but—and this is a big but—the number is trending up at an alarming rate. What will happen to traditional brick-and-mortar retail and wholesale distribution channels when this number hits 30 percent?

Before you answer, it might be instructive to think about the model for diapers. com and Toys R Us. Both companies sell diapers, and both companies sell them at or near their cost. "Loss leader" is the retail term of art for selling a product at or below your cost to draw people into your establishment. In both cases, the goal is the same: Lure the customer in with low prices on a commodity item, and get them to purchase additional items (that you make a profit on) while they are there.

This is a well-tested retail strategy. It works every time.

Now, imagine a generation of Millennials[5] coming of age and needing diapers for their new babies. They are 100 percent digital native and highly mobile and smartphone-centric. When a significant number of these young parents start to subscribe to commodity products online, they will cease to visit brick-and-mortar retail establishments as frequently as they did before the technology changed their behaviors. To make matters worse, the early adopters will be the smartest, most affluent quintile of the population. Nothing is going to stop this trend. Are you prepared for it?

How will you lead your organization into a brave new world where the biggest of the big-box stores lose up to 30 percent of foot traffic through their doors?

Disintermediation by online subscription of commodity goods comes in many forms. One of my favorite examples is dollarshaveclub.com. It offers many lessons for digital leadership in a connected world, so let's look at a few of them.

DollarShaveClub.com: Changing Retail One Blade at a Time

First, DollarShaveClub.com identified an opportunity: Razor blades are expensive, and there is a market for a good blade at a good price. Razor blades are so expensive that to succeed, DollarShaveClub.com would not need to offer the least expensive blades, just a good balance of convenience, quality and price.

If you haven't seen DollarShaveClub.com's promotional video: http://www.youtube.com/watch?v=ZUG9qYTJMMsI&feature=youtu.be, take a minute to watch it. It is the very definition of what a well-produced, well-targeted video should look like.

---

5 Generation Y, also known as the Millennial Generation, is the demographic cohort following Generation X. There are no precise dates for when Generation Y starts and ends. Commentators use beginning birth dates from the later 1970s, or the early 1980s to the early 2000s decade. "Generation Y." Wikipedia. Wikimedia Foundation, 23 Oct. 2012. Web. 27 Oct. 2012. <http://en.wikipedia.org/wiki/Generation_Y>.

If you haven't been to DollarShaveClub.com, take a minute to visit the site and see how a transactional website can be seamlessly merged with an experiential branded website. This is very hard to do, and dollarshaveclub.com has accomplished it nicely.

Of course I had to test it out, so I signed up, and within a week, I received my first order. There are three packages to choose from. I chose the middle one, a four-blade razor. For $6 per month, DollarShaveClub.com sends you four new blades each month.

The package was simple and thrifty. In the envelope were a handle, four blades and some promotional material.

Just to level set for a second, beards are as individual as the people who have them. So my experience is probably not typical, but it might be. **Spoiler alert:** The following paragraph about my personal shaving experiences may fall into the category of TMI (too much information), so you may want to skip it.

I have a pretty tough beard and have been a happy The Art of Shaving customer for over a year. I use the entire system: pre-shave oil, shaving soap applied by brush and a five-blade razor (the one that is spoofed in the dollarshaveclub.com video), the Gillette Fusion blade. The eight pack of these blades costs $25 at The Art of Shaving. That's $3.13 per blade plus tax and shipping if you order online, or plus tax and transportation if you buy them at the store (DollarShaveClub.com's four-blade blades cost $1.50 each, no tax, and shipping is free).

I did the side-by-side comparison test, and I can tell you without hesitation that the Gillette Fusion blade absolutely smokes the dollarshaveclub.com blades. It is no contest. A two-week-old Fusion blade will give you a closer shave than a brand new DollarShaveClub.com blade, hands down. So while dollarshaveclub.com founder Michael Dubin asks the rhetorical question, "Are our blades good?" then answers, "No, they're f#@king great!" that answer might be both self-serving and a bit of an exaggeration. As far as my beard is concerned, the DollarShaveClub.com blade offers a suboptimal shaving experience.

I am a little sad that I can't save $6 per month on blades. It would have been fun. But for less than the cost of a trip through the Midtown Tunnel, I'm not going to sacrifice a great shave with The Art of Shaving's Gillette Fusion blades for a truly bad shave with DollarShaveClub.com blades.

OK, back to the future. Let's not lose sight of the impact of subscriptions on traditional retail businesses just because dollarshaveclub.com doesn't offer a superior product. Maybe other men will find the shave excellent or acceptable for the money. And as we all know, if dollarshaveclub.com is not the ultimate solution to this market inefficiency, fiftycentshaveclub.com or some other new entity will figure out how to do it better and the business will evolve.

One very instructive and interesting factoid is that dollarshaveclub.com is not the least expensive way to purchase blades online. It is confirmed that dollarshaveclub.com buys at least some of its blades from a company called Dorco. At <u>dorcousa.com</u> you will get an up-close, personal look at how badly a website can be designed, how impossible it can be for a consumer to make a comparison purchase and how, no matter how much they offer to save you, your time is worth more. Dorco's website is neither experiential nor transactional; it's just bad.

You'll learn more from comparing dollarshaveclub.com and dorcousa.com than you will from trying to figure out how to do a viral video like dollarshaveclub.com's for your company.

Mike, I wanted to love dollarshaveclub.com. I think you guys are geniuses. I just wish the blades actually were f@#king great.

Transparency vs. Information Asymmetry

The asymmetry of knowledge (where one party has more knowledge than the other) has been the distinguishing characteristic of profitable transactions for all of recorded history. Actually, I'm sure it predates recorded history; I just can't prove it.

Economists have long studied the imbalance of power caused by information asymmetry and have nomenclature for special kinds of examples. After you Wiki "information asymmetry," check out "adverse selection" and "moral hazard." Now, just browse around Wiki and Google and you will assimilate the knowledge contained in two semesters of undergraduate Econ classes in less than 10 minutes—oh, wait! That's what this chapter is all about.Among the benefits of living and working in a connected world, there is one that stands out as "the" change agent for humanity:

virtually instant access to information. Now, I have said time and time again that information is not knowledge, but in the hands of someone who can contextualize it, the speed of information is directly equated to economic success.

The simplest of the information asymmetry problems has many names. Some like to call it "price transparency" or simply "transparency." But that oversimplifies and undervalues the problem. Just knowing the wholesale price or cost price of an item does not always help you get the best deal, although most sellers would consider it a challenge to constantly deal with consumers who knew their margins.

The bigger problem is that the underlying engine of all transactions is now imperiled at scale by instant access to information. To make matters worse, the technology that makes all of this possible is trending upward at an exponential rate. Doing business is about to change in a fundamental way, and there's nothing anyone can do to stop it.

Transparency is powerful. In the very near future, sellers will not be able to overcharge for nonessential and non-time-sensitive goods and services.

You may push back and say that the auto industry has been dealing with this since the advent of the Net and that they have survived the onslaught of commodity pricing. You may also think that online auto sellers and brick-and-mortar auto sellers have reached equilibrium. This is patently false. The sticker price on a car has nothing to do with the price the dealer is paying. Neither does the invoice price. If it did, there wouldn't be any car dealers. Do you really think you can run a floor-planned[6] car showroom making $25 per transaction? Seriously?

Showrooming

On the other hand, when you are looking at a Samsung 46" HDTV Model number UN46D7000LFXZA at an electronics retailer, then take out your handheld device and search the Web, you will instantly find it at a significantly lower price.

---

6 A form of financing pertaining specifically to inventory. A lender will purchase the inventory from the borrower, and as the inventory sells, the borrower will repay the debt. It is essential that the creditworthiness of both parties is established and that a procedure for if the inventory does not sell is in place before the lending takes place. "Floor Planning." Definition. N.p., n.d. Web. <http://www.investopedia.com/terms/f/floor-planning.asp>.

You know it is exactly the same unit you are looking at in the showroom. Can the brick-and-mortar retailer match the online price? Will it? As this problem, known as "showrooming," gets bigger, retailers are going to be highly de-incentivized to purchase inventory they cannot sell at a profit.

If retailers don't stock items, manufacturers will have to perfect just-in-time inventory. When they do (many already have), what will become of the post-4G retail store? How will the retail environment have to adapt to be relevant in a world where everyone who walks in has instant access to the lowest price, free shipping and no tax from an online vendor? Buyers already know more than sellers about the features and benefits of products—the Web is the perfect tool for that. What does the retail store or showroom have to offer? How will manufacturers adapt production and finance to accommodate the change?

Now, let's expand this idea to every other business we can think of. As we move toward transparency, retail transactions certainly get tougher, but service businesses get hit just as hard.

If the job of service professionals is to transfer the value of their intellectual property into wealth, how much will transparency hurt?

If I have made a living knowing more than my clients know, I'm in trouble. If I do deals by bringing my clients to third parties who add value, transparency takes away my value (since lists of my preferred vendors or partners, case studies and clients are readily available). Do I really need to hire a private wealth manager to figure out who the best hedge funds are? Not any more. Do I need to hire a consultant to tell me who the most interesting start-ups are? Nope. Do I need to hire an agency to get me the best talent? Of course not. All the metrics and information I need to arrive at a business decision are at my fingertips. Transparency is going to strengthen us in ways that traditional competition has not, or it will destroy us.

Is knowledge power? No. It really never has been. Information asymmetry in a transactional environment is power. And now, thanks to transparency, information asymmetry is becoming harder to create. To survive, the basic structure of transactions will have to adapt.

What Does It Mean to Own Something in the 21st Century?

Free vs. Paid: The Wrong Debate

There are only three business models: I pay, you pay or someone else pays. That's it. "I pay" means that I (the publisher of the content) am willing to fund the creation, production and distribution of the content for my own purposes. "You pay" means that you are willing to pay me for my content. "Someone else pays" means that a third party is willing to pay me so that you can consume my content. Some of my KPMG friends have pointed out that from the content producer or publisher's point of view, there are really only two models: I pay or I get paid. I like to include the idea of third-party involvement, because it is so common to the media business.

In practice, we see three adaptations of the three models: Ad supported (broadcast), Subscription/Pay per View (premium content) and the dual revenue combination of ad support and subscription (cable/print, etc.)

The common thread to all of these approaches to getting paid is the transfer of the currency of "cash" from one party to another for good and valuable consideration. Again, I pay, you pay or someone else pays—cash, credit card or check (with two forms of ID), please.

This is how business has been done for as long as anyone can remember. But it has very little to do with how business needs to be done now.

Cash is the currency we are all most familiar with, but it is far from the only one. Information is a currency. Knowledge (curated and filtered information) is a currency. Imagine a car service driver on West 38th Street in Midtown Manhattan driving aimlessly, waiting for his dispatcher to send him a job while riding around slowly looking for a gypsy-style fare. If he had a device in his car (such as Uber7) that told him

---

7 uber.com and its associated app provide licensed, professional drivers the ability to receive and fulfill on-demand car service reservations as your private driver.

that there was a fare waiting on 38th and Lex, he could almost immediately translate that information into cash. If several cars had that information (and they do; Uber has a huge fleet in New York City), it would be less valuable to the driver but still give him a significant competitive advantage over other, less technologically evolved car service drivers. If every car service used Uber, it would simply be a business methodology with only commodity value. Any Wall Streeter will tell you that information (especially exclusive information) is cash. But a housewife living in the suburbs with knowledge of a sale at a local store or with an electronic promotional offering can translate that information into cash just as easily.

The key concept here is that a contemporary business model needs to "translate" the value of one currency into another. In the example above, the currency of knowledge is translated into currency of cash: You can't get cash directly by distributing the information; it must be translated.

Information is an easy one. How about the currency of attention? Attention is such an established currency that it is written into the grammar of the English language: You "pay" and "receive" attention. In the online world, there are a couple of other currencies that are key: intention, fame, passion, respect/street cred.

Google translates the currency of "intention" into about $50 billion a year. Google knows that when you come to Google, your "intention" is to find something. Google delights in telling you that in .02 seconds, it has found 2,374,345 things that you may have intended to click on. Google offers all of its services for free. Why? It knows that if it pays off your intention with valuable information, you will keep coming back. Like any good casino, Google assumes that if you show up often enough, you might click on something Google gets paid for. It's the best currency of intention to cash money translation engine ever created.

I could spend a few hundred pages explaining how one might translate the value of the currencies of fame, passion or respect/street cred into the currency of cash. But I think you get the idea. Is it about promotional videos or selling t-shirts, concert tickets and merchandise? No, of course it isn't. But a holistic approach to the doing of business is often an excellent way to translate disparate currencies into cash.

Will this work for a commodity like emergent or local news? It depends upon the originality and quality (in the eyes of the consumer, not your definition of quality) of the content. No one is going to pay for something they can get free of charge or easily do without. Newspapers filled with yesterday's news delivered to the outside of your house aren't a good product in the 21st century. A wireless electronic delivery system that notifies me of relevant, emergent news is more in keeping with informational needs of today's consumers—oh, wait, that's what I get on my smartphone and online.

Can a free online product ever make money? Yes, absolutely! You simply need to understand the currencies of the Internet and build models that translate the value of those currencies into wealth.

Can you frame this conversation as free vs. paid? No. Not if you are trying to get someone to pay you cash directly for something that is ubiquitously available for free. Free vs. paid is not the great debate. It's a no-brainer: Free wins! Valueless vs. valuable, scarce vs. ubiquitous, demanding of attention vs. commanding of attention are the debates, and the winners will be the individuals and organizations that can most effectively translate the value of content into wealth.

Ancient Akko and the Apple App Store

I had occasion to visit the ancient city of Akko, which is located on a promontory at the northern end of Haifa Bay in Israel. Akko's history dates back to biblical times, and if you are a student of history, it's worth a quick trip to Wiki to learn about it.

Today, Akko is more of a tourist attraction than a working port, but while walking through the bazaar in the center of town, I could not help but think about its modern-day equivalent, the Apple App Store.

I remember an old economics professor once lecturing about perfect competition, which is a theoretical condition where no participant influences the price of the product that is bought or sold. This particular professor used an Arabian bazaar as an example of where perfect competition might exist. Some of the conditions for this theoretical marketplace include:

- A large number of buyers and sellers who are ready, willing and able to buy and sell at a certain price.
- Very low or no barriers to entering or exiting the marketplace.
- Prices and quality of goods are known to all participants.
- Very low or near-zero cost of transactions on either side.
- Homogenous product characteristics: small or no variation in the attributes or characteristics of any product in the marketplace.

There are other conditions required to enable economics students to create pricing models using perfect competition as a benchmark, but they really won't add much to this discussion.

As I walked through the bazaar in Akko, I noticed several merchants selling similar merchandise. Spice merchants, produce merchants, souvenir merchants, jewelry, shoes, etc. There were several, maybe a dozen or more, of each type of stand. From the perfect competition model my old professor described, I would have assumed that the prices for a spice like cumin or paprika would have been identical or at least very, very close from merchant to merchant. And from the same example, I would have expected prices for a replica of the Ark of the Covenant (knocked off from the Raiders of the Lost Ark movie, made in China and sold to tourists as authentic local merchandise) to be similarly priced.

Not even close. Prices varied by what the prospective customer was wearing, whom the customer was with (alone or with a licensed tour guide or from a cruise ship), how aggressive the prospective buyer bargained and several other tangible factors. Stated prices were only a guideline, and actual sales prices were absolutely all over the map. Astounding.

Here in a veritable sea of identical merchandise, packed in so close that you could hear a deal being made at the competitor's shop, prices varied wildly. So wildly that I had to try it out for myself. I purchased a golf hat at one stand for 25 shekels—10 shekels under the asking price. Not twenty steps away, I purchased the identical hat for 15 shekels by just pointing to the competitor's shop—not a word spoken by me. I purchased the identical hat half a block away for 23 shekels, but I got a small bag of figs to go with it for "free." So much for perfect competition.

Which brings me to my favorite modern-day Arabian bazaar, the Apple App Store. There you can find the same WiFi signal monitoring app for free, $.99, $2.99 and $4.99 and even higher prices. They all do exactly the same thing on the same devices. They all take advantage of Apple's iOS and they all work fine (otherwise Apple would not approve them for sale). Why are they different prices? How can you tell them apart? How should they be marketed? How are they not commoditized?

Almost every class and kind of app in the Apple App Store is sold under conditions that fit the model for perfect competition. But, like the Arabian bazaar in Akko, the laboratory conditions do not really exist. Instead we see a vibrant marketplace with a true *caveat emptor* theme. Truthfully, I was tickled to see a physical example of the virtual App Store here in Akko. It's not a way that most Americans shop today, but I guess everything old truly is new again.

Why Can't I Pay You?

I hate being a criminal. It's really no fun at all. I bet you hate it too. But the amount of casual piracy that is woven into the fabric of our connected lives is really starting to bother me.

One of my good friends sent me an email recently. He just returned from a weeklong trip and wanted me to hear this "new, great song" that he downloaded from a friend while he was away. I have no idea what song it is. It has a cryptic filename, no song name, no artist's name, no metadata. It is truly a pirated work because without any identification it has no promotional value. I had no luck with the usual music identification services, and the worst part was that there was no way to pay for it, even if I wanted to.

This got me thinking about the plethora of new handheld devices. All of these devices have a MAC address (Media Access Control Address), which is basically a globally unique number assigned to each device by the manufacturer. Technically, your computer does too. But the handhelds are generally sold by networks with a contract for network access.

The difference is profound. You can't use your Verizon phone on the AT&T network unless you contact both companies and they switch the service for you. In practice, any compatible device will work on any compatible network, but the network will require a billing relationship with the device before you use it. As I said, when you use your computer to access the Internet, you do not need permission from any network owner; you just need access to the Internet. (Note: While AT&T's 3G network is GSM/EDGE and Verizon's is CDMA/EVO, there are several dual-system devices on the market, and both companies use LTE (Long-Term Evolution) as their 4G standard.)

This should not come as news to you, but stating the obvious points out a nontrivial difference between the open Internet and the evolving wireless broadband universe. The telephone companies (carriers) have been expert in extracting every single cent from every possible billable feature on every device. This is because they own your access to content from end to end. (The term of art is a "Walled Garden." It's beautiful on the inside, but a prison nonetheless.) Is that the future? If I am a content provider, must I sell through a carrier to make a sale in the mobile broadband world?

If so, what happens when I get an email like the one I just received from my friend? It has a file attached that I have no way to purchase. How might we solve this problem? Do we need the carriers to get involved? They have a billing relationship with each user. In theory, the carriers could bill for content if they knew it was supposed to be charged for.

Of course, we might invent a tool that enabled content owners to identify their work and a small piece of computer code that would identify copyrighted material and offer the prospective user an option to purchase it. Such a button would actually allow us complete unrelated micro-transactions on any connected device. It would certainly help keep honest people honest.

Who gets paid and how? is always the number one question in the meetings I attend. With the advent and explosive growth of identifiable broadband devices, it's really time to work on some systems that will allow us to seriously reduce casual piracy. No one wants to be a criminal, and most people who can afford smartphones and broadband plans are willing to pay fairly for what they consume.

As a digital leader in a connected world, it's up to you to make this work. If digital transactions are super easy, it is relatively easy to get people to digitally transact. If you make it impossible to digitally transact, the results should not surprise you. Just hope that your business can withstand the loss of good customers who would have been willing to pay you, if only you'd given them a convenient way.

What If Video Were Data?

Let's do a thought experiment to help us visualize the future of commercial message management. Of course message management is just a way to describe the way content gets to a consumer. We have a bunch of other names for it: television, radio and newspapers, to name a few. Traditionally, these media have been used to manage messages to consumers, and most of them (with the exception of broadcast network television and some of its affiliates) have been enjoying dual revenue models (advertising and subscription).

This idea of maintaining a dual revenue stream is the topic of all kinds of conversations wherever cable operators and broadcast television stations cross paths. Everybody is interested in what's going to happen when free content over the public Internet breaks the knees of the cable industry and cripples the broadcasters, etc. I can easily argue both sides of this, but for now, let's try to imagine a profitable future, a future with real upside potential where technology helps us make money, not the other way around. Here we go!

Imagine that your video content is data. Not bits and bytes, data. How would you distribute it profitably? For example: The news anchor is reading a story to the camera. There are supporting graphics and text. What if all of it were in a database and all of it were made available all of the time via a feed to the outside world? Call it a broadcast feed, if that metaphor helps you. But imagine it not as video, but as packets of data.

Carry this metaphor as far as you can. All of the content you broadcast or print or distribute is now just packets of data that are being broadcast out to the world. This feed carries all of your branding, all of your unique filtering, all of the things that make your content yours, but instead of its normal form, it's data.

OK, now let's have some fun. Head over to the Apple Apps Store if you have an iDevice, or if you don't, go to the Facebook applications area and have a look at all of the things that third parties are doing with other people's data. There is an immense amount of value being created and quite a bit of currency changing hands in both places. Super-smart (and not-so-smart) and super-talented (and not-so-talented) companies and individuals are creating ways to assimilate, aggregate, utilize, synthesize and profit from other people's data. It is awe inspiring. The Apple Apps Store has some programs for $4.99 and then similar programs for $.99 and then hundreds of knock-offs that are free. Each is competing in an open marketplace for dominance, each with its own consumer value proposition—like I said, awe inspiring.

Now, imagine your television, radio or print content transformed into branded data and distributed throughout the universe as a trusted source of excellence. After all, that's what your brand is. Imagine giving all of the third-party application makers and users a way to consume your content (data), while you concentrated on making it and feeding it out there. Imagine that your advertisers want their data associated with your data, because your data are trusted and stand for excellence. Of course, data are extremely measurable (because they actually are bits and bytes), and all of a sudden it becomes clear—we are now in the super-digital age. Here virtual DMAs (Nielsen Designated Market Areas) replace geographic ones. Third parties, who live inside these communities of interest, create applications for the hundreds of millions of registered social network users, not the other way around. We are remunerated because our data drive sales and enhance brand experience for our associated advertisers, and our costs decrease with every entity that lends a helping hand. A pseudo-dual revenue stream is maintained because we can charge (or share revenue with) the programmers who use our data.

Does this model work for every type of content? No, of course not. But it's perfect for emergent content such as news, weather, sports and live events or events that are topical and that audiences believe are live.

In very short order, we (in the traditional media business) are going to have to transform ourselves into massive branded data spigots; there is no other choice. We will have to adopt Tom Sawyer's fence-painting paradigm, if only because it is too expensive to create all of the workflows and deliverables that a non-interoperable, consumer-driven world requires for scale.

I love this plan. It does require a whole bunch of stuff to change. But the state of our economy has certainly accelerated that change. And to that end, it has unburdened us with the need to preserve the status quo. It's crunch time, and no matter how much it hurts, we can all transition to the Information Age and keep our messages managed and sponsored. We just need to get everyone to help us paint the fence.

Piracy Is Rampant – It May Always Be

Why We Had to Stop the Stop Online Piracy Act, and PIPA Too

Choosing to not understand how the Internet works is no longer an acceptable option, especially for those tasked with regulating it. The late Ted Stevens made Congress's ignorance of the World Wide Web famous when he said the Internet isn't a dump truck, but a series of tubes. This statement was made in 2006, but its replays on YouTube echo a lack of understanding that was brought into the spotlight by the war over SOPA and PIPA.

SOPA and PIPA are acronyms for pieces of legislation that were proposed by Congress, created with the hope of finally bringing some sort of regulation for copyright infringement to the Web. The proposed bills wanted to stop websites from spreading copyrighted material or selling counterfeit items. More than that, the laws would have affected websites like YouTube, where users often post content that violates the rules in the terms of service. If one of these videos slipped through the cracks, technically the government could require ISPs (Internet Service Providers) and search engines to block the site, which was not really a workable, or even reasonable, solution.

The bill debated in the House of Representatives was SOPA, the Stop Online Piracy Act. Spearheaded by Republican Representative Lamar Smith, the proposed legislation came under fierce scrutiny from tech industry monoliths like Google and Facebook, as well as security experts who thought it was technically disastrous. PIPA, the Protect IP Act, was the Senate's even weaker version of the bill.

The "You're destroying the Internet!" versus "We need to stop these digital thieves and protect intellectual property by passing a sweeping law!" skirmish played out for a while mostly outside of the public eye until some of the biggest websites put the issue front and center. Wikipedia, Google, Mozilla, Wired and Reddit were among the sites that used their valuable digital screen real estate to protest what they considered egregiously ignorant legislation.

These protests seem to have been effective. Google announced that it acquired over 8 million unique signatures on its SOPA/PIPA petition, just from a link on its homepage. Wikipedia directed over 7 million people to their local representatives so they could express their opposition to the legislation. But what exactly was being opposed?

One of the controversial provisions would have given the government broad power to make service providers block websites at the DNS (Domain Name System) level. Many agree that this would not only be ineffective, it could have posed potentially serious security issues. If legislation is ever to be passed to regulate content on the Internet, much greater technical foresight will be required.

A less technical provision that is ruffled feathers would have made the maximum penalty for streaming or downloading copyrighted content five years in prison. Most reasonable people would agree that a mother caught illegally downloading a torrent file of Josh Groban's Christmas album probably should not face the possibility of five years in jail.

As flawed as the language and provisions of SOPA and PIPA were, there was one part of both bills that made absolute sense: Stealing content is wrong.

Sadly, there is a fairly large group of people who believe everything on the Internet should be free. Hacktivists routinely cripple websites of businesses, organizations and even governments that oppose the idyllic "free Web." While I completely understand how post-Napster digital natives have come to believe that stealing copyrighted material (and depriving rightsholders of the value of their intellectual property) is OK, I violently disagree with their thesis.

This put me in a very strange position. I was aligned with people seeking the same short-term result I was seeking: These bills, as written, could not become law. But what would we do after SOPA and PIPA were defeated? That's easy—we are going to have to figure out how to reconcile the fact that "we the people" who respect copyrights and "we the people" who don't are going to have to work together to solve this problem. It won't be easy. In fact, it may never happen.

Which raises the question, "How should intellectual property rights be protected?" Unfortunately, the technological answer is far, far easier than the political one. Before we can protect content creators, we need to agree that content needs to be protected. Considering the passionate beliefs on both sides, I'm not optimistic.

A lack of optimism is highly uncharacteristic of me. In almost every case, I believe that technology and education can solve this kind of problem. What this impasse means in practice is overwhelmingly bad for creative people and the businesses that profit from their intellectual property and overwhelmingly good for people who think ownership of content is unfair.

Pop culture arts may cease to be a business in the next 10 years or so. If you can't protect potential upside profits for an investment as inherently risky as an artistic endeavor, you are unlikely to make the investment in the first place.

Is there a reasonable, profitable business model for content creators and distributors in the 21st century? As a digital leader, it's up to you to help find it.

Cyber-Security and Cyber-Warfare

Cyber-Security Starts with Strong Passwords

We are under attack 24/7/365, and there is nothing we can do to stop it. Hackers, organized crime, hactivists, hobbyists and precocious teenagers are working day and night trying to hack into everything of value. As I have emphasized throughout this writing, information is currency, so you should think of hackers across the full spectrum from digital muggers to digital genocidal maniacs to digital war criminals.

We can't do much to protect ourselves against attacks on big institutions. When Facebook or Google get hacked, we're going to be hacked too. But for personal and professional vigilance, cyber-security begins at home.

Depending upon where you live, you probably lock your doors and windows when you leave your physical house. If you consider your neighborhood to be a little dangerous, you may even leave your doors and windows locked when you are at home. To succeed in a connected world, you need to keep your digital doors and windows locked all the time. So after you read this section, it will be time to stop reading and change all of your passwords. It is the metaphorical equivalent of locking your windows and doors, and it is really not optional.

Here's how: First, make you passwords strong. I am an advocate of strong passwords—inconvenient, long, strong passwords. 7-1d7w!Ka was my Yahoo! password until Yahoo! got hacked and I was forced to change it. Again, when an institution like Yahoo! gets hacked and your passwords are stolen, it doesn't matter how strong they are; you need to change them. But strong passwords will protect you from most direct attacks by casual hackers exactly the same way locking your glass doors and glass windows on your physical house will protect you against casual burglars and robbers.

So, as I was saying, 7-1d7w!Ka is an old password of mine. Can you guess the phrase I based it on? Hint: It's written in LEET, and it is a famous phrase from the 1939 movie classic The Wizard of Oz. Got it?

7-1d7w!Ka is an abbreviation for, "Toto, I don't think we're in Kansas anymore." The letter T is represented by the number 7. The uppercase letter I is represented by a 1. The lowercase letter i is represented by an ! and the other letters are just letters.

"Toto, I don't think we're in Kansas anymore" gets shortened to T-IdtwiKa, which gets translated to LEET as 7-1d7w!Ka, which is about as strong of a password as you can create, and it's very, very easy to remember.

Here's a simple LEET table. Try to make a few long, strong passwords by picking a favorite phrase or quote from a movie or book and using the first letters of each word to construct your password.

A	@	4	^	/\	/-\	aye			
B	8	6	13	|3	/3	ß	P>	|:	
C	©	¢	<	[	(	{			
D	)	|)	[)	?	|>	|o			
E	3	&	€	Ë	[-				
F	f	|=	/=	|#	ph				
G	6	9	&	C-	(_+	gee			
H	#	}{	||	]-[	[-]	)-(	(-)	/-/	
I	1	!	¡	|	]	eye			
J	]	¿	_|	/	</	(/			
K	X	|<	|{	|(					
L	|	1	£	|_	1_	¬			
M	|v|	|V|	/\/\	(v)	/|\	//.	^^	em	
N	|\\|	/\/	[\]	<\>	/V	^/			
O	0	()	[]	°	oh				
P	¶	|*	|o	|°	"	|>	9	|7	|^(o)
Q	9	0_	()_	(_,)	<|				
R	2	®	/2	12	l2	l2	|^	|?	lz
S	5	$	§	Z	es				
T	7	+	†	-|-	']['				
U	µ	|_|	(_)	L|	v				
V	\/	^							
W	VV	\/\/	\\'	'//	\\|/	\^/	(n)		
X	%	*	><	}{	)(	ecks			
Y	¥	J	'/	J					
Z	2	7_	~/_	>_	%				

Making very strong, inconvenient passwords and using them is one of the best things you can do to protect yourself against casual hackers.

That said, we all have dozens of websites that we visit, and it is really not a brilliant idea to use the same password for all of them. You can do it, but it increases the risk that one good hack will give you a serious headache.

There are two programs I like that solve this problem. One is free, but a little geeky. The other costs money, but works like a charm. KeePass (Windows) and KeePass X (Mac) are free, open source password managers. And 1Password is a very nicely packaged, paid solution that will let you automatically create and manage a large number of extremely long, strong, cryptic passwords on all of your devices: Windows, Mac, iOS, Android, etc.

The value of this kind of password management software is that not only can it help you create excellent passwords and autofill them for you, it can help you change your passwords very quickly.

The more we put our lives in the cloud, the more vulnerable we are. Getting a handle on password management is a best practice requirement for success in a connected world. So check out some password management software and get a system in place (mobile device passwords too). And please don't write you new password on a sticky note and put it on your monitor.

Weapons of Mass Disruption: The Super-Cyber-Wars Have Begun

While we were busy worrying about Kim Kardashian, sound bite politics and banning 64-oz cups of sugary soft drinks, the first super-cyber-weapon was quietly spying on and attacking things it didn't like. Huge, stealthy computer viruses with capabilities that sound like the stuff of science fiction have been hard at work, collecting data and data and more data, waiting to strike. It is clear that we live in a world at war, cyber-war ... and this is just the beginning.

In May 2012, while they were busy on assignment from the International

Telecommunication Union (ITU) and looking for something else entirely, Kaspersky Lab discovered Worm. Win32.Flame—the world's first super-cyber-weapon. Kaspersky named it "Flame." What made Flame a super-cyber-weapon? Kaspersky Lab said it was highly sophisticated and malicious with complexity and functionality exceeding all other cyber-weapons known to that date. Scary stuff.

Kaspersky went on to say that while "the features of Flame differ compared with those of previous notable cyber-weapons such as Duqu and Stuxnet, the geography of attacks, use of specific software vulnerabilities, and the fact that only selected computers are being targeted all indicate that Flame belongs to the same category." It was just more powerful, harder to detect, harder to understand, bigger and nastier than anything they'd ever seen.

As best as anyone could tell, Flame was "designed to carry out cyber-espionage. It can steal valuable information, including but not limited to computer display contents, information about targeted systems, stored files, contact data and even audio conversations."

To add to everyone's anxiety, Kaspersky engineers think that Flame may have been deployed as early as March 2010. Why wasn't it detected until May 2012? Welcome to the nature of highly complex, highly targeted super-cyber-weapons.

But this got more interesting. Kaspersky Lab also posited that Flame was so complex and sophisticated that it must have been created by a government entity. And as if to validate that assumption, shortly after the Kaspersky announcement, Moshe Yaalon, Israel's vice prime minister and minister of strategic affairs, said, "For anyone who sees the Iranian threat as significant, it is reasonable that he would take different steps, including these, in order to hobble it." He went on to say, "Israel is blessed with being a country that is technologically rich, and these tools open up all sorts of possibilities for us."

You're probably wondering how this will impact us here in the United States. That's the wrong question. The right question is: "How will this impact you?" Here's why.

Most kinds of malware are colloquially referred to as "computer viruses." Virus is the metaphor because this kind of computer program exhibits some of the characteristics of biological viruses. Viruses aren't exactly alive, but they can procreate, replicate, mutate and die. Computer viruses are very, very similar, so the metaphor is a good one.

As you know, almost every civilized society on earth has banned the use of chemical weapons. Aside from their cruelty, the simple fact is that, once deployed, biological viruses don't discriminate. They can't tell friend from foe, right from wrong or good from evil – those exposed become infected. Yes, the metaphor of a virus is excellent for super-cyber-weapons.

Although the creators of Flame (whoever they are) did not do as good a job as they could have, Flame was discovered. Its discovery raises the questions, How many more super-cyber-weapons are out there? and Who and what are the targets?

By far the most disturbing attribute of super-cyber-weapons is that intellectual property of this type is neither proprietary nor controllable. You can ban all the cyber-weapons you like, criminalize them, stigmatize their usage, make all the noise you want – they can't be stopped until they are discovered. This is an intellectual arms race with zero barriers to entry. You don't need enriched uranium, you don't need controlled substances, you don't need any contraband, you don't need anything but a laptop and a Starbucks card. (The Starbucks card is for Internet access. Don't start thinking that cyber-terrorists hang out at Starbucks—although they might.)Who are likely targets? Everyone and everything of value, such as power plants, air traffic control computers, banking computers, digital medical records—and that's just the stuff that's easy to imagine. Corporate cyber-espionage is going to become pandemic. In the age of Big Data, information—especially private, proprietary information—is a prime target.

Is anyone safe? No. Is any computer immune? No. Is there anything we can do? Yes. Get the best virus protection you can afford and keep it updated. Purchase cyber-security insurance and comply with the guidelines so your computers and data are as safe as is practical. Use common sense and go long on cyber-insurance companies and antivirus companies. (This is not financial advice. I'm just trying to be cute, although both cyber-insurance and cyber-security will be growth industries over the next 20 years.)

Commenting on uncovering Flame, Eugene Kaspersky, CEO and co-founder of Kaspersky Lab, said: "The risk of cyber warfare has been one of the most serious topics in the field of information security for several years now. Stuxnet and Duqu belonged to a single chain of attacks, which raised cyberwar-related concerns worldwide. The Flame malware looks to be another phase in this war, and it's important to understand that such cyber weapons can easily be used against any country. Unlike with conventional warfare, the more developed countries are actually the most vulnerable in this case."

Cyber-Terrorism vs. Cyber-Warfare: Defending the United Networks of America

Seven-year-old Mark Fielding looked up from his computer. He was very annoyed. "Mommmm!" He yelled in a way that was sure to get her attention. "The Internet is down again." It was the last thing she heard before the lights went off. Mark turned on his smartphone and opened a blank browser window to use as a flashlight. He found his mother by the front door. She was looking out on a darkened landscape. Neither of them had any idea of just how dark it truly was.

Ten minutes earlier, a remarkably powerful computer virus had destroyed six of America's most important data centers. Five minutes earlier, a different piece of code killed every caching server on the three biggest CDN's. At Zero Hour, the attack culminated in the computer-controlled destruction of the entire power grid in North America. It would take days to fix and months to fully repair, and the cost would be measured in trillions, not billions.

America's days as an economic superpower had ended. All the financial data at the IRS was destroyed. Six of America's major financial institutions could not access their records. No one could find a digitized medical record in any database with proper metadata (the data that describe data). With our data destroyed, our economy ceased to be. The breakdown of social services was immediate and devastating. The doing of life would never be the same. America, as we knew it, was gone.

Who did this? The Chinese? The Russians? Religious Extremists? No. It was a small group of unaffiliated, highly motivated computer hackers. Whom did they work for? Anyone. Who paid them? No one. Why did they do it? Because they could. What was their punishment? Sadly, they were never found.

What an unsatisfying way to end a great science fiction story. No enemy? No villain? No narrative? Try selling it to Hollywood. Such bad writing!

Perhaps, but this is not science fiction. This is a very real probable future for the

United Networks of America. Which is where we all live right now!

Most people interface with the United Networks of America through the world wide web. However, you can also gain access through your wireless phone or over the public Internet. You may think of the Net as Google, Yahoo, eBay, Amazon or CNN.com, but there are literally millions of private, local area and wide area networks that all have access points on Al Gore's Information Superhighway. These networks contain all of the information that describes us. I called it Metamerica earlier. And if you remember, metadata are data that describe other data, and Metamerica is the information that describes us.

In the Information Age, America without Metamerica would be practically useless. Where is Metamerica? As we've previosuly discussed, it is in the data centers at Google, the IRS, our banks and financial institutions, medical facilities, business networks and even our home computers. And for all practical purposes, it is unprotected and unprotectable.

This fact alone should be enough to scare any thinking person. But I have not yet begun to describe the hard part of the problems we are facing.

What is a war? The dictionary says it's an "armed conflict between nations." The dictionary does not say what they have to be armed with. What is terrorism? What is a crime?

In the Information Age, what is a country? What is a state? What is a nation? What is a tribe? What is a community of interest? What is an enemy? Where do the enemies live? Do they need to be people?

What are weapons? What are military targets? What are civilian targets?

According to a June 2012 Quadrennial Defense Review review by the U.S. Department of Defense, "On any given day, there are as many as 7 million DoD (Department of Defense) computers and telecommunications tools in use in 88 countries using thousands of war-fighting and support applications. The number of potential vulnerabilities, therefore, is staggering."

"Moreover, the speed of cyber attacks and the anonymity of cyberspace greatly favor the offence. This advantage is growing as hacker tools become cheaper and easier to employ by adversaries whose skills are growing in sophistication."

Defensive measures have already begun. The Pentagon created US Cyber Command. But how will we know when we are being attacked by a country, an enemy, a terrorist, a criminal, a mob, a gang, an individual? When would the military know it was supposed to get into the fight? CIA? NSA? FBI? Google Security? A consortium of concerned citizens with antivirus software on steroids? How can you tell an invasion from a teen-age prank?

William Lynn, former U.S. deputy defense secretary, described the cyber challenge as unprecedented. "Once the province of nations, the ability to destroy via cyber now also rests in the hands of small groups and individuals: from terrorist groups to organized crime, hackers to industrial spies to foreign intelligence services. This is not some future threat. The cyber threat is here today, it is here now," Lynn said.

I spend a fair amount of my time counseling clients on how to deal with the fact that we are at the midpoint of the analog-to-digital transition. Today, as I am fond of saying, we have analog leaders and digital citizens, analog commanders and digital soldiers. If the pen is mightier than the sword, the "digital pen" is mightier than a million ballpoints! Forget the threat of cyber-attacks for a second and think about how an enemy might use the power of social networking and the ability to instantly publish any type of message globally to its advantage.

It's time to rethink the public Internet, computer networks and the infrastructure of our digital world. The currency of information is as important and valuable to our economic sovereignty as tangible stores of value. Bits of gold dust or bits of information—in the super-digital age, they deserve equal protection.

Cyber-Warfare: Fighting and Winning

Will the United States be a superpower, a cold warrior, a sovereign nation, a first-world entity or a third-world entity in the 21st century?

Before we can get into the strategy and tactics of fighting and winning a cyber-war, it would be helpful to understand who we are and whom we are fighting.

Are fans citizens?

That may sound like a strange question. But in a connected world, fans self-select into communities of interest. When a community of interest forms around a pop-culture icon, a sports team or a movie, we call the members of the community fans. When a community of interest forms around a political or religious worldview, what should we call them?

Obviously, the tyranny of geography does not apply to a connected world. Global communities of interest can form around any topic in a very short time frame. Look at any trending topic on Google or Twitter to get a better appreciation of the speed of information in the Information Age.

The colloquial association with terms such as "community of interest" or "fans" is that of a passionate but casual affinity toward a particular subject. While this may be true in the traditional sense of the words, one disturbing trend has been the zealousness and vitriol of passionate, sometimes only partially informed, advocates of particular worldviews and their remarkable ability to voice their opinions as facts.

Internet purists will tell you that the system is self-correcting and that "facts" and "fact checking" are actually overly scrutinized in online settings. There is considerable evidence to the contrary. Real facts are hard to come by in the Information Age—there's too much noise surrounding them.

One interesting consequence of the Information Age is the ability for people to cocoon themselves in the information they agree with and wrap themselves in the security of hearing only what they want to hear. This becomes more and more important as communities of interest metamorphose into social media trust circles. (A trust circle is simply a group of people whose opinions you trust above other sources.) In the Information Age, trust circles not only self-assemble, they are among the most powerful forces we face.

In the advertising and marketing business, we used to complain about the decentralization of mass media in the United States (truthfully, people are still complaining about it, but you can't put the toothpaste back in the tube). As everyone loves to remind us, back in the mid-20th century, there were three networks and you could inform, enlighten and entertain (or brainwash or propagandize) a remarkably large percentage of the population by disseminating information from only a few sources.

After the advent of the cable television industry and a new technology called the electronic remote control, the enemy was the fragmentation of the audience—you needed more tools to reach the same number of people in more places.

In the past few years, we've gone from a 500-channel universe to a multi-million-channel universe. Consumers (viewers) can no longer be described as fragmented; they are atomized. They have not gone away; they are simply self-assembled into millions of overlapping trust circles. This trend of consumers taking control of their information consumption and distribution will continue as long as the technology progresses. In other words, everything will continue to decentralize at an accelerated rate for the foreseeable future.

Whom are we fighting? Is it a nation, or a self-assembled mist of atomized, like-minded individuals? Is the enemy made up of virtual nation-states?

If the very definition of government is an "empowered central command," certain questions are now unavoidable. For example, what popular currencies do governments use to govern?

Cash. Our cash is backed by the full faith and credit of the United States of America.

Military power. Why do people believe in our cash? Could it have something to do with the 10 Nimitz-class supercarriers on active duty around the world? It might. Gunboat diplomacy is well understood and understandable. Our conventional military power is so extraordinary, no thinking nation would attack us with conventional military force. Any version of traditional warfare would be met with such overwhelming force, I don't have enough hyperbole to describe it.

Information. America is the world's grandest experiment in freedom of expression. As we know, the control of information is directly translatable into cash or military power.

When I discussed these three currencies of government with a rather well-known economist, I was told that cash and information are equivalent for this argument. Since our nation's wealth is only supported by the belief in our posterity, information or propaganda—choose your own word to describe the Tao of the people – information and the military are the two most powerful currencies in the Information Age.

According to my friend, they are symbiotic. I'm not sure I agree, but I'm not an economist. Assuming that information and the military are two sides of the modern coin of the realm, one must ask: What constitutes weapons-grade information in the Information Age?

- Top secrets (Governmental, corporate, personal, etc.)
- Data (all of our records: financial, medical, consumption, etc.)
- Metadata (descriptions of data such as people's identities, financial data, etc.)
- Network topology (our digital infrastructure)
- Telecommunications networks
- Access to the power grid
- Access to access points in the networks

I have had the remarkable pleasure of speaking with several high-ranking military officials about this and the issue of analog leaders and digital soldiers. I was told that practically every military leader (of sufficient rank) in the United States Armed Forces is an expert in the strategy and tactics of wars fought on battlefields. That is very comforting. But the most devastating wars we are likely to fight in this century will not be fought on battlefields. We are going to fight cyber-wars, several of them, and they are going to target our economic sovereignty in ways that conventional wars never have.

Awhile back, hackers targeted Google and a couple of dozen other West Coast tech companies. The attacks were specific, vicious and successful. The NSA, CIA, FBI, TSA, Homeland Security folks, Army, Secret Service—name your governmental agency or arm—had no idea. There were no air raid sirens, no red alert Klaxons. The nation did not know it was under attack. It was.

If you don't know the history of Google, it is very well retold in Ken Auletta's book Googled: The End of the World as We Know It. In the book, you will find a description of what Google is. I'm sure you think of it as a search engine, and if you are more enlightened, you may know about its other products and ad-supported businesses. People search for information on Google hundreds of billions of times per month, and Google has a copy of every search ever entered into a Google search box. It learns from every search, and it is optimized to deliver the best, most relevant advertising to the user based upon that search. Don't be fooled into thinking that an advertising company can't possibly have national security value. Information is "the" currency of the Information Age, and Google has a 100 percent search information monopoly. No other entity on earth comes close.

After the attacks, there was much Sturm und Drang about who did what to whom. Was it an attack by the Chinese government or just a couple of unaffiliated hackers? If it was a nation attacking us, how would we know, and how would we fight back?

Testifying before the Senate Intelligence Committee, the top U.S. intelligence official warned that U.S. critical infrastructure is "severely threatened" and called the cyber-attack on Google "a wake-up call to those who have not taken this problem seriously."

"Sensitive information is stolen daily from both government and private sector networks, undermining confidence in our information systems, and in the very information these systems were intended to convey," said Dennis C. Blair, director of National Intelligence, in prepared remarks outlining the U.S. intelligence community's annual assessment of threats.

After the attack, Secretary of State Clinton said, "A new information curtain is descending across much of the world," as she called the growing Internet curbs the modern equivalent of the Berlin Wall. She went on to say, "We stand for a single Internet where all of humanity has equal access to knowledge and ideas," as she cited China, Iran, Saudi Arabia and Egypt among countries that censored the Internet or harassed bloggers.

Sadly for us and for Secretary of State Clinton, bureaucracy and diplomacy are not going to get this done. The U.S. government is practically powerless in this arena. This was not a conventional attack. There were no enemy combatants, no bombers, no nuclear missiles. This was a cyber-attack with a specific target. Could there be a more asymmetrical warfare problem? A few unidentifiable, highly skilled, highly motivated individuals against the United States of America. Just how many Nimitz-class supercarriers would you like to send, and where might you send them? Just for the sake of argument, should we consider nationalizing Google? The only way to punish a nation-state in the Information Age is to cut off its access to information. A combination private- and government-crafted information isolation is the economic equivalent of destroying the Ancient Library of Alexandria for any specific country. If China wants to play an Information Age game of schoolyard name calling, let's cut off its access to information. It's a level of economic sanction that we could not accomplish any other way.

Obviously, multinational corporations will have a huge problem with this. So will everyone else. It's a war, and when you are at war, people get hurt! That's why you try to avoid wars. But we can't fool around with this. We have analog leaders who think in analog ways, and they are being asked to deal with a remarkably complex set of digital infrastructure issues that, honestly, only a very few people truly understand.

Obviously, we can't nationalize Google, but I've made my point. The only way to fight a cyber-war is with cyber-tools. We need a bunch of them, and we need them fast! To fight and win a war in the Information Age, we need to control the information. Control that. In many ways, Google already has.

Cyber-War 1.0: China vs. the USA—We're in Trouble!

"In the past year, one in seven large organisations detected hackers within their systems." This is the highest level recorded, said the recently released PwC 2012 Information Security Breaches Survey. It was completed in conjunction with Infosecurity Europe and supported by the Department for Business, Innovation and Skills. The survey goes on to say, "This year's results show that security breaches remain at historically high levels, costing UK plc billions of pounds every year." The additional summary stats are compelling as well:

The average large organization faces a significant outsider attack every week, small businesses one a month.

20 percent of organizations spend less than 1 percent of their IT budget on information security.

Customer impersonation is up threefold since 2008. Financial services are affected most.

This survey was Eurocentric, but the stats were similar all over the free world. Between good, old-fashioned hackers and anonymous and other self-described "hacktivist" groups, the world has become a much more dangerous place.

When bad guys attack businesses, the results can range from inconvenient to very expensive to completely disruptive, but what will happen when the country gets attacked? What will Cyber-War 1.0 look like?

China is home to some of the most dangerous hackers in the world. Although the Chinese government denies any formal ties to any hacker groups, most experts agree that the level of sophistication demonstrated by Chinese hackers betrays that denial. It would surprise no one to learn that these groups are funded and trained by a government that can conscript its best and brightest citizens.

Here in the USA, our cyber-defense is lumpy. Our best, most digital businesses are well protected by the smartest cyber-defenders money can buy. If you're wondering how that's possible, Google brags that it employs more PhD's than any other organization on earth. America's high-tech community is second to none.

Sadly, this extraordinary private cyber-army does not work for nor is it conscripted to protect our country or its population. It is employed by corporations to protect corporate assets.

Where does our federal cyber-army come from? Obviously, municipal agencies and the military do not have the kind of HR budgets or stock incentive plans that high-tech companies are famous for.

Let's review. I've got Chinese hackers, who are government sponsored, highly paid (in relative terms) and fully incentivized vs. American civil servants, whose digital skill sets perfectly positioned them for government work.

We're doomed. What do I mean by doomed? You know, Armageddon, end-of-days, extinction-level-event doomed.

Sometime soon, some hacker group (probably from China) is going to hit us hard. Maybe 20 million Americans will wake up one day and find their checkbook balances at zero. Maybe the power grid in the Northeast will go dark. Perhaps we will find ourselves with our credit reports altered or our credit card bills in disarray.

If the hackers are really smart, they may wait until too many people have too much sensitive data in cloud storage facilities and take them out. If you can think of it, so can a hacker, and I can assure you, the outcome will not be good.

Right about now, you're probably wondering why I have chosen to write such an alarmist, fear-mongering, sensationalist message about cyber-warfare. The answer is simple: Very few people are paying attention to this subject. Cyber-security is a lot like air: You don't miss it until it's gone.

What can you do? First and foremost, get into the subject. Do what you can to understand what your best practice business continuity plan should look like. Start with the question, "What will happen if ..." and keep the dialog going. Involve your C-suite, your tech guys and your customers.

Start a dialog with your colleagues. How will your company function under the stress of digital disruption? Where is the breaking point? When do you switch over to backup systems? Who is in charge of the decisions?

In the aftermath of most terrorist attacks, someone from some agency goes on CNN and tells everyone how they knew it was coming and how everyone should have prepared for it. Why wait? You can prepare now, and trust me, the economics of your organization will benefit from the planning.

What will the outcome of Cyber-War 1.0 be? If we're ready, the vast majority of us will never know it happened.

Digital Leaders and a 21st-Century Workforce

Could You Get a Job Today?

Do you have community management skills? Can you set up and man listening posts? Are you an expert at setting up and processing Google Alerts? Can you clean up, size and manipulate digital pictures and graphics? Are you a PowerPoint ninja? Do you have more than half of the PC keyboard macros for Excel under your fingers? Can you write an SQL query? Can you craft custom reports in salesforce? Do you have expertise in a particular kind of CRM software? Can you interpret and respond to questions regarding Google Analytics? Are you facile with FTP software? Are you a master of digital communication in your industry?

These are just a few of the questions you might field in a job interview this year. I recently listed a job opening for an administrative assistant, and to be honest, I am appalled at the lack of understanding of how to apply for a job, let alone what might be required to obtain one.

As a digital leader, you can decide how many of these guidelines you agree with, but if you are a digital worker looking for a job, here are a few tips for applying for one in the Information Age.

Cover letters matter. Your cover letter should be in pure text and in the body of an email. No fancy fonts, no images, just text. The topic sentence should be awesome and separate you from the pack. The supporting paragraph should make me want to hire you without looking at your résumé. It must, must, must mention the things your prospective employer is seeking and describe why you are the perfect candidate. Proofread this document several times. "I lernt frm xperience that i'm a realy grate receptionist." is an actual sentence from an actual cover letter. I have no idea what this person's résumé looked like; I just copied the sentence for this book and deleted the email.

Résumés matter. Take the time to craft the résumé for the job you are applying for. If you haven't worked in the industry before, say it in the cover letter and say why you think your experience will apply. If you have worked in the industry, take a moment and figure out what your résumé should look like for this opportunity. Résumés should be .pdf files—do not send Word documents or .txt files or PowerPoint documents or anything other than a one-page (two-page max) .pdf file. The only exception to this rule is when an HR department specifically requests a Word file. They do this so they can input the data into one of several automated systems that parse Word files. If this is the case, make sure your document is using a compatible Word format—.doc vs .docx may make all the difference with legacy HR document processing software. Read and understand the digital submission requirements. If you can't get that part right, your fate is sealed.

Honesty matters. Don't put "Expert in Microsoft Office" on your résumé if you are just "proficient." During our telephone interview, I will ask you a question that an expert can answer, and when you can't, you're out. I have no time for people who cannot do honest self-assessments of their capabilities.

Skills matter. This is the Information Age. You need Information Age skills. Yes, you will learn a great deal on the job, but you need to come to the opportunity with very high-level digital skills. Why? Because there are literally a dozen digitally skilled candidates who will apply for this position. They are more cost-effective for me to hire, because they can do more for the same money I will have to pay you.

Work ethic matters. I want people around me who are self-starters and who know that the sentence "Can I help you?" is the least helpful sentence you can utter. What's the right way to impress me? "Shelly, I've identified this issue. I have three solutions. Please tell me which one you would like me implement." I will do anything for people who approach work in this manner—they are awesome!

Understand what work is. If you are looking for a skilled job, understand what work is—a mechanism to translate the value of your intellectual property into wealth. This is a nontrivial distinction between a "job for a paycheck" and a career. If you want a job, you are not someone I want to hire for a full-time position. If you have a career and you are looking to grow by acquiring knowledge, tempering it with wisdom and forging it with failure, I want you on my team!

Understand the value of what you know. There's an old cliché, "Youth is wasted on the young." Today, when you're looking for a job, don't waste the value of your youth. Yes, you may be young and inexperienced, but you have a valuable asset in your age. If you are born after 1989, you are a digital native. This means that you think differently, act differently and in fact are different than the middle-aged hiring manager you are speaking with. Your inexperience and youth are also a liability. Get smart and use this combination of strength and weakness to your advantage. Our culture aspires to be young—this is news you can use.

What if you don't have the necessary skills? This is the key to everyone's future. You must acquire them. No one can afford to hide behind the affectation "Digital is for kids." It's nonsense, and it is a virtual guarantee that you are unemployable in the 21st century. You no longer have the luxury of saying it. In fact, you cannot even think it. Social media are being used to "Occupy" places and overthrow governments. If you're not a social media expert, you are at a strict disadvantage. Facebook and LinkedIn (and 500 other social networks) are replacing email. Google is mapping the interiors of retail stores. Amazon is giving people $5 off of any purchase made by taking a picture of an item in a brick-and-mortar store and then making the purchase via your mobile device. There is no more analog; the world is digital. More to the point, there are now only two kinds of people and two kinds of devices: connected and not connected.

Job One. I'm still looking for an administrative assistant with awesome digital skills to work for my executive admin. Will we find the right person? Of course we will. For all of the horrible résumés and cover letters submitted, there were several gems. But the sheer volume of worthless communication from unemployable candidates has been remarkable. If job creation is our number one national priority, maybe we should start by helping people learn how to properly prepare for employment in the Information Age and then teach some basic job-hunting skills.

Basic Digital Literacy: What Is Really Required

Are you employable? If you want to be, you will need to have and continually hone some basic digital skills. Here's a quick overview as articulated in my book *Overcoming The Digital Divide: How to Use Social Media and Digital Tools to Reinvent Yourself and Your Career.*

Google. There is no way around this one; you must be great at using Google. You must know how to search using specialized parameters. You should be able to find only results published in the past 24 hours, who is linking to your recent blog post, all the sites that published your recent press quote, exact and entire phrases (not just keywords), all the Web pages that include your name but not your company name and many other advanced search techniques. Visit http://www.google.com/insidesearch/ to see what Google thinks you should know.

Microsoft Office. Specifically, this includes Word, Excel and PowerPoint. If you don't understand how to use Track Changes in Word, create formulas in Excel or set up slide transitions in PowerPoint, you're in trouble. Take the time to learn these programs, by searching online for tutorials, using the program's Help function or simple trial and error or taking a class. Understanding these programs is nonnegotiable for the modern businessperson.

Basic HTML. You don't need to be able to code a website from scratch, but you should have a basic familiarity with the computer coding language HTML. Can you determine what keywords your competitors are using in the View Source function on your Web browser? Can you italicize, bold and underline the comments you make on Web blogs? Can you converse intelligently with your Web design and development team about basic needs and tasks? Can you use an "href" to place an image on a webpage? If not, I promise you, there are dozens of upstarts fresh out of college who can. This information is easy to find online, and you can teach yourself in a few hours.

Basic photo and video editing. Images and videos are undeniably powerful forms of communication. They push any piece of digital content to the next level. They demand attention. As such, you must be able to manipulate them, even if only in the most basic manner. For still photos, either with a full-featured program like Photoshop or a free program like Picasa or GIMP, you should be able to crop, resize, add text, adjust resolutions, remove blemishes, add filtered effects and generally improve the quality of any image. For video, you don't need to be able to re-create Michael Bay–inspired special effects or put a story together with the artfulness of Steven Spielberg, but you should be able to cut together basic highlights, add text slate and generally pace video material that will satisfy and engage its audience. For the vast majority of people, Apple's iMovie or Windows' Movie Maker will more than suffice for basic video editing needs. Conveniently, these come prepackaged with either operating system, and there are about a billion videos on YouTube showing you how to do exactly what you want.

Social media. This is also a nonnegotiable requirement for doing business in the 21st century. Do you have a Facebook profile? If not, go sign up now. It is extraordinarily easy. I mentioned this in the previous section, and it is so important, I'll mention it again. Just visit http://www.facebook.com. Follow the instructions. Next, do exactly the same thing at http://www.twitter.com and http://www.linkedin.com. None of this is optional. Just do it!

There are literally dozens of other skills you can add to this list, but this short list will get you started. It is helpful to know that almost anything you need information about can be found by searching online. Google, YouTube and Wikipedia are great places to start.

Lastly, this is much easier than it sounds. So don't worry about it. Like I said, just do it!

The Attack of the Pridefully Ignorant

I had to take a one-day trip to Boca Raton, Florida, to attend a family event. During my trip, I met several people (of a certain age) who feigned interest in my profession. I was drawn into conversation after conversation in which I had to defend the existential necessity of digital literacy. Would it be cliché for me to tell you how many of these individuals had flip phones? Would it be stereotypical to describe the number of doctors, lawyers and retired investors who have their secretaries print out their emails? Would it be hackneyed to recount the pridefully ignorant way that all these individuals espoused the reasons they lead an unconnected life? Let's give it a shot.

An attorney who has a remarkably successful practice in South Florida told me that he doesn't see any reason to follow the industry trend of hiring electronic discovery experts. He boasted to me that remaining antiquated protected his practice from modern invasive electronic discovery techniques. He went on to tell me how he knew all about this "tech stuff," but it just wasn't important enough for him to invest in. I pointed out that we were in the Information Age and that practically everyone who communicated did so using digital tools. About five seconds into my response, I just changed the subject; I am not prepared to argue with the pridefully ignorant.

I have about ten other examples of this kind of insanity, but I'm sure you get the point. So if you are willing to think about overcoming the digital divide, let's go over a few key points.

First and foremost, inject yourself into the process. If you want to become more digitally literate as a means of enhancing your ability to transfer the value of your intellectual property into wealth, you must dive in. How? Start by listening.

Do you have a Facebook profile? If not, go sign up now. It is extraordinarily easy. If you are daunted by the task, screw your courage to the sticking place and click this link: http://www.facebook.com Follow the instructions. If you can read, you can get this done in less than 10 minutes. Don't worry about your privacy settings right now. You're not going to do anything on Facebook today that will compromise your privacy or open you up to identity theft. I promise.

Once you have a Facebook profile, start sending friend requests to your actual friends. Resist all temptation to make it a popularity contest—just invite people you know well. And only friend people you know well. Once you've got a bunch of Facebook friends, start listening. Forget about your wall and your profile page and just watch the news feed. It will only take a few days for you to start understanding what Facebook should (and should not) be used for in your community.

Want to get more into social media? Join some groups. There are Facebook groups on almost every subject you can think of. Join, and just listen. There's no need to post anything until you are ready.

Next, do exactly the same thing with Twitter. Set up a profile page, start following people you know and people you want to know and work with and just listen. It is the fastest way to become digitally literate in the world of social media.

If you want to interact with people on Twitter, consider replying to their tweets instead of just tweeting stuff out. It changes the dynamic of Twitter and will make you an instant part of the community.

The world is bifurcated. There are only two types of people and two types of devices: connected and not connected. The mantra of the pridefully ignorant is: "Digital is for kids!" If you wish to be pridefully ignorant, keep saying it. You will soon fade into complete unemployability and communicative irrelevance.

To lead a connected life, you need to be connected. This means having a smartphone and learning to use it. If you really don't want a smartphone, get a tablet or a high-end color e-reader and carry it with you everywhere. You will need a device to be connected to the Internet, and you can't connect without a device. Get one!

Not a smartphone or a tablet person? Tough! You need to be. So get with the program. The only way to make this leap is to make it.

How will you know what gear to buy? It doesn't matter what you get as long as you get something. iPhone, Android—I don't care. You won't care either, at least not now. There will come a time when you will care. At that point, you will make another purchase and you won't need anyone's advice about what it will be.

Lastly, make a resolution to learn how to use some keyboard shortcuts and some digital productivity tools. It could be as simple as forcing yourself to use all of the Microsoft Word keyboard shortcuts for formatting, or as adventurous as installing Text Expander (Mac) or Phrase Express (PC) to enhance your word processing efficacy. Like I said, the only way to become digitally literate is to inject yourself into the process, and enhanced productivity is a big step toward that commitment.

Although I was brutally attacked by a horde of pridefully ignorant technophobes in Boca Raton, I escaped. I hope you will too.

You Don't Know What You Don't Know

"You don't know what you don't know" is one of my favorite phrases. It's an admonition I take seriously. No matter how hard you study, no matter how much knowledge you acquire, no matter how much wisdom you possess, there is always more to learn.

Digital communication is here to stay. Intel said that at the end of 2012, there were just over two billion people connected to the Internet. It is projecting three billion by the end of 2015 and hopes the world gets to four billion by the end of 2020. Intel has a selfish reason for this prediction: It makes the chips for the billions of devices that will support this connectivity.

Cisco says that by 2015 there will be approximately 15 billion connected devices in the world. That is an incredibly large number. And 100 percent of these devices are digital. While it's true that some of the radio signals that are used with these devices are analog, there are computers doing all of the heavy lifting. So for all intents and purposes, the world of communication is 100 percent digital. Which raises the question, "Why would anyone even suggest that digital literacy is an unimportant skill in the 21st century?"

Did Tiger Woods know that he was making a digital audio recording on a remote server? That bit of digital illiteracy cost him a lot. Did Eliot Spitzer know that a wire transfer was actually a digital file transfer and that, while private, it was not anonymous? Obviously not, and that lack of digital knowledge cost him big. How about Anthony Weiner's and Brett Favre's experiences with digital photos? Their collective lack of knowledge about simple digital file transfers yielded unfortunate consequences too.

Do you know how well your online presence matches your offline presence? When I Google you, will I find what you would expect me to find? How do you look on LinkedIn? Do you actually believe that there is a hiring manager in the connected world who will not check you out on Google, Facebook, Twitter and LinkedIn before considering you for a position? The reasons to become digitally literate are endless.

I am not advocating digital communication over in-person connections. In fact, I'm not advocating anything other than that the "pridefully ignorant anti-21st-century communications tools group" should consider that they don't know what they don't know.

These are early days. Social media are in their infancy, and we have no way of knowing whether Facebook or Twitter or any one of the other 500 popular social networks will be around in a few years. What we do know is that while some people are using social media for egocasting, others are community organizing and still others are overthrowing governments. Same tools, different applications.

Here's the key take-away: If you are not connected, there is a significant conversation going on about you in a place you know nothing about, care nothing about and don't believe exists. Like I said, you don't know what you don't know.

Science, Technology, Engineering and Math Really Matter

The Higgs Boson: What We Can Learn

There are several different ways to study the nature of matter, but one of the most fun is to smash very tiny particles into each other at extraordinary speeds and see what happens.

This is what CERN, the European Organization for Nuclear Research, does with its Large Hadron Collider (LHC), one of the largest, most complex machines ever built. This $10 billion proton beam smasher on the Swiss-French border was built for one purpose: to determine if the Higgs boson actually exists. What is a Higgs boson? Great question. Go to Wikipedia and read about it.

Particle physics is not generally a pop-culture vocation. But circa July 4, 2012, particle physicists were all over the news. It seems that there is a 99.99 percent chance that experiments at CERN had given us our first look at a particle that, according to a spokesperson, "is consistent with a Higgs boson as is needed for The Standard Model."

Try to contain your excitement.

The Higgs boson is considered, by a significant number of mainstream physicists, to be the "missing link" to the "Standard Model" of particle physics. According to Wikipedia, "The Standard Model" of particle physics is a theory concerning the electromagnetic, weak, and strong nuclear interactions, which mediate the dynamics of the known subatomic particles." So this discovery is exciting because confirming the existence of this predicted particle helps to validate the Standard Model.

But even if this new particle is not a Higgs boson, there is still good news. It may be the first of a new type of massive subatomic particle that will completely alter our understanding of the universe. So the news was win/win for scientists everywhere.

That said, scientists are pragmatic, so after saying "Thanks, nature!" Fabiola

Gianotti, one of the CERN team leaders, told reporters, "The Standard Model is not complete" but "the dream is to find an ultimate theory that explains everything—we are far from that."

There is no practical use for a Higgs boson, and we are not any closer to understanding gravity, or dark energy, or a unified theory of everything, so what have we learned from this 17-mile-long, $10 billion atom-smasher?

Well, first, we learned that there is a huge value to pure scientific research. Science for its own sake can take us on a journey toward the answers to mankind's oldest and most basic questions. And although none of the scientists I spoke to would even venture a guess about the practical use of this discovery, I can't help thinking that it is both inspirational and aspirational.

How many kids will start asking questions like, "What is everything made of?" or "Why do we need a theory of everything?" How many adults will be inspired to learn about the forces of nature and the scientific method? How many of us will aspire to learn all we can learn about the world around us and our place in the universe? I love these questions and the thousands more that will come from this remarkable discovery.

I also love the complexity of this particular problem and the technology, mathematics and computer science needed to solve it. We may invent new kinds of math to help with this search, build new kinds of computers, rethink materials handling, change the nature of manufacturing or learn to do a thousand other things from the science needed to do this kind of research. It's simply awesome!

Get excited about this. Read about it. Tell your friends about it. Think about it. Who are we? Why are we here? What are we made of? Why do things around us behave the way they do? What is our ultimate fate? What can we do to change the world? There are so many more questions you can ask ... so ask! Striving to understand who and what we are may be our highest, best purpose.

Man vs. Watson: The IBM Jeopardy Challenge

In 1877, Thomas Edison invented the phonograph. Ever the showman, Edison liked to demonstrate his device by allowing people to speak into the machine and then playing the recording back for them. However, back in the late 1800s, the technology was outside almost everyone's conceptual understanding. Up to that point in history, the only thing that could mimic the sound of one's voice was a ventriloquist, so people thought it had to be a trick. Clergymen came to pronounce it "the devil's work" and to discredit Edison.

But here's the really fun part. Edison used to charge people 25 cents to try to "fool the machine." A person who spoke Latin (a dead language) would speak Latin into it, and of course it would speak Latin back to the person. People wondered how Edison was able to teach a machine to speak Latin. A person would speak Chinese into the machine, and it would speak Chinese back to them. Again, people would wonder how the "Wizard of Menlo Park" taught the machine to speak Chinese.

People simply did not understand the concept of a recording. And back then, no one had ever heard a recording of his or her own voice.

Arthur C. Clarke once said, "Any sufficiently advanced technology is indistinguishable from magic." Can you imagine seeing a technology and completely misunderstanding what it was doing and how it might work?

For most people, that's what happened on February 16, 2011. That's the evening that a very sophisticated IBM computer named Watson beat two human opponents in a friendly game of Jeopardy.

The match was broadcast over two days giving the blogosphere, the tweetisphere and Facebookistan time to light up with pro-Watson and anti-Watson sentiment. In almost every case the arguments went something like this:

Moron 1: "Machines will never be better than people!"

Moron 2: "Yes they will!"

Or,

Metaphysicist: "Machines will never fall in love."

Physicist: "Neither will I. So what?"

Or,

Theologian: "Machines will never achieve consciousness and will never have souls."

Atheist: "They do not need consciousness. Simple self-awareness will do, and no one has a soul."

Or,

Creationist: "If G-d intended man to fly, he would have given him wings."

Evolutionist: "We have been inventing tools to make our lives better since the beginning of history. Watson is simply another tool that we have created with our G-d-given brains!"

If you spend a few minutes and take even a cursory look through the Interweb, you will find an extraordinary amount of writing by passionate parties on both sides of the John Henry argument. (If you don't get the reference to John Henry, he was a "steel driving man," a mythological 19th-century folk hero who collapsed and died after winning a race with a steam-powered hammer. Good story, and there's a pretty good song that goes with it too.)

Sadly, it's the wrong argument and absolutely the wrong way to look at the event.

First, kudos to IBM and the producers of Jeopardy for crafting a fantastic and wonderfully entertaining demonstration. This technology could have been displayed in an infinite number of less interesting ways. The TV show was a remarkable mixture of showmanship and science reminiscent of the great science exhibitions of the 19th century.

If you're a student of innovation or just want to learn about how important showmanship was to industrial age innovators, do a little research about the World's Columbian Exposition, aka the Chicago World's Fair of 1893. Thomas Edison and

Nikola Tesla duked it out for the right (and the contract) to light up the fairgrounds with electric lights. Although Tesla and Westinghouse ultimately won, the real winners were science and innovation. For the first time in history, people saw (in a dramatic fashion) how electric light bulbs were going to change our world.

It is in this spirit that we should look at the IBM Jeopardy Challenge. And it is in this spirit that we should applaud Watson. This technology represents American know-how and American ingenuity and clearly demonstrated the power and promise of corporate research and development departments. This was a wonderful demonstration of the nexus of pure scientific research and applied engineering. Watson represents the highest, best purpose of our educational system in America: to relentlessly explore the boundaries of our imaginations and to seek answers to questions we do not even yet know how to ask.

We live in the Information Age, and information is the most valuable currency of our time. If the United States is to remain a superpower, it needs to be a digital superpower and an information superpower.

Equally importantly, there are more honors students in China and more honors students in India than there are students in the United States. We need to inspire our children to become world-class experts in the arts and sciences. Brilliant young minds will ensure our posterity.

You may think that Watson is a machine and that because it has a name and a voice, you can anthropomorphize it. Don't go there. It's the wrong road. Watson is a humanized, show biz–ized demonstration of technology that will keep us safe and secure in a cyber-war. It is a clear path to better educational tools, better medical tools, better weather prediction tools, better business tools and tools that will help us succeed in a connected world.

Watson and its offspring are in a class of machines that inspire us while they inform, enlighten and entertain. To those who think that Watson and similar technology dehumanizes us, shame on you for being fooled by the machine.

Conclusion

Privacy, security, commerce, community, communication, healthcare, entertainment —no matter which field you choose, technological progress is the story of mankind. We cannot separate ourselves from our inventions. They reflect who we are and what we believe. We build tools to help us do things—good and bad. And if we are to make a smooth transition from the Industrial Age to the Information Age, we are going to need excellent digital leaders to guide us. It is my sincere hope that this book has both inspired you and empowered you to think about how you will lead your business to prosperity in the Information Age.

My parents were both music educators, and from a very early age, I was taught that "practice does not make perfect; perfect practice makes perfect." The lesson was practical and profound. If you allow yourself to play wrong notes when you practice, you will allow yourself to play wrong notes when you perform.

The rigor of perfect practice—playing the same thing over and over again until it is 100 percent flawless—borders on boring. Music is funny that way. You can make a typo when you write a book (I'm sure one or two typos in this writing got by our proofreaders and editors), and readers point them out but generally don't judge the quality of the writing by the technique of spelling or punctuation.

But if a musical performer plays even one wrong note out of 100, everyone knows and judgments are made. Professional musicians play 100/100 perfect notes. If professional musicians played with the accuracy of the best baseball players who ever lived, you would not pay to hear them play. Could you imagine a concert where only 3.5 to 4 out of every 10 notes were played correctly? It would be comical, to say the least.

As I became more and more musically proficient, it occurred to me that "technical superiority was the key to artistic freedom." The more technically proficient I became, the better I could musically express myself, until one day, if I could hear it in my head, I could play it for others to hear.

This concept of technical superiority being the key to artistic freedom is applicable to everything in our connected life. As best as I can tell, technical superiority is the key to all freedom—the freedom to create, the freedom to communicate and the freedom to lead in a connected world.

The goal of this book is to give you a way to think strategically about how work and life are changing. Now it's your turn. Go deep, inject yourself into the process and practice perfectly to digitally lead and succeed in our connected world.

---

i   Kurzweil, Ray. The Singularity Is Near: When Humans Transcend Biology. New York: Viking, 2005. Print.
ii   "How Would You Like to Manage Your Twitter Account, Today?" Tweepi V2.0. N.p., n.d. Web. <http://tweepi.com/>.
iii "Urban Legends." About.com. N.p., n.d. Web. <http://urbanlegends.about.com/>.
iv   "Snopes.com: Urban Legends Reference Pages." Snopes.com: Urban Legends Reference Pages. N.p., n.d. Web. <http://www.snopes.com/>.
v   Washington, William Yardley; Kitty Bennett Contributed Reporting From. "Palin Start: Politics Not as Usual." The New York Times. The New York Times, 03 Sept. 2008. Web. <http://www.nytimes.com/2008/09/03/us/politics/03wasilla.html?_r=3>.

INDEX

A

advertising
  local, 84–6
  online, 16
  targeted and addressable, 66–7
  types of, 17
AIM buddies, 29
Akko (Israel), 94–6
Amazon, 82, 109
Android, 81, 124
anonymity, 60–61
antivirus companies, 107
Apple, 73–4
  App Store, 94, 96, 99
Arab Spring, 60
AT&T, 97
Auletta, Ken, 113

B

bandwidth, 13
Blair, Dennis C., 114
blogs, 26
  blogmobs, 29–31
  blogosphere, 31, 58
  censorship of, 114
  microblogging sites, 31
Bloomberg News, 63
Born Digitals, 6
bots, 31–3
brand/lifestyle advertising, 17
Breitbart, Andrew, 55
Bush, George W., 25

C

cable television, 112
call-to-action advertising, 17
Cameraphone, 36
CERN, 127–8
China, hacking by, 115–17
Cisco Systems, 2, 9
civil liberties, 30
Clarke, Arthur C., 129
Clinton, Hillary, 114
cloud storage systems, 3–4, 70–71, 74–5
CNN, 24–5, 109
Colbert, Stephen, 52, 54
Comcast, 11–12
"Common Sense" (Payne), 61
community management, 118
community of interest, 111
computer viruses, 105–108
Concepcion, John, 56–7
Consumer Privacy Bill of Rights, 76–7
consumer relationship management (CRM),
  81–2
content management systems (CMS), 66, 81
Cook, Cheryl, 55
cover letters, 118
CRM software, 118
customer relationship management (CRM),
  65
customer service, 44–6
cyber-espionage, 106–107
cyber-insurance companies, 107
cyber-security, 103–105, 117
cyber-warfare, 115–17

Made in the USA
Lexington, KY
27 November 2013